9/06

Recycle The Essential Guide

With an introduction by Lucy Siegle

Black Dog Publishing

How to use this book

The following pages will tell you everything you've always wanted to know but were too afraid to ask about recycling. The body of the book is divided into two interleaved sections: materials and case studies.

The materials sections are divided into the main categories of recyclable goods: paper, glass, metal, plastic, compost and household waste. This last category contains within it all those products that don't quite fit into your green box; computers, wood, textiles, paint, tyres and batteries. Each of the sections will explain what the material is, how it is sourced, what the recycling process entails and what you can do to recycle more. Interspersed amongst these sections, in the coloured boxes marked with a ♻ you can find a variety of little stories, facts and suggestions, and at the end, a materials directory will point you on to the organisations in your country that can provide further information.

The case studies are slotted into the materials section, and pay tribute to exceptionally successful or innovative approaches to recycling, as well as representing the darker side of the subject – that of the illegal export of waste. From small local schemes such as those in London's borough of Hackney, to national efforts, such as that of Green Dot Germany, we hope that these case studies will give you a sense of what the imperatives are, and what the future approaches to recycling may be.

When reading these case studies, it is important to view them in the context of the country's policy and economic position. Rich countries have a special responsibility for global environmental stewardship, both because they have the capital to invest in new technologies and because they are the primary producers of waste and carbon dioxide emissions. Recycling is as crucial in developing countries as it is elsewhere. However, poorer societies tend to have a different approach to waste – after all, when resources are scarce, anything is fair game, and rubbish is often re-used in the most inspired, imaginative ways.

The directories at the end of the book are divided into three sections: design, governmental and non-governmental. The first promotes eco-design – highlighting the importance of buying goods made of recycled products. The latter two will point you on to organisations that can help you take further action and can provide further information about recycling schemes in your area.

With ever decreasing space for landfill, and climate change tangibly evident, there is a rising awareness of the responsibility we hold for the welfare of our planet. We hope this book will give you a starting point for action, and illustrate the imperative of the three 'R's – Reduce, Reuse, Recycle.

Contents

Case studies

ORDEM E PROGRESSO

Guide to case studies from around the world

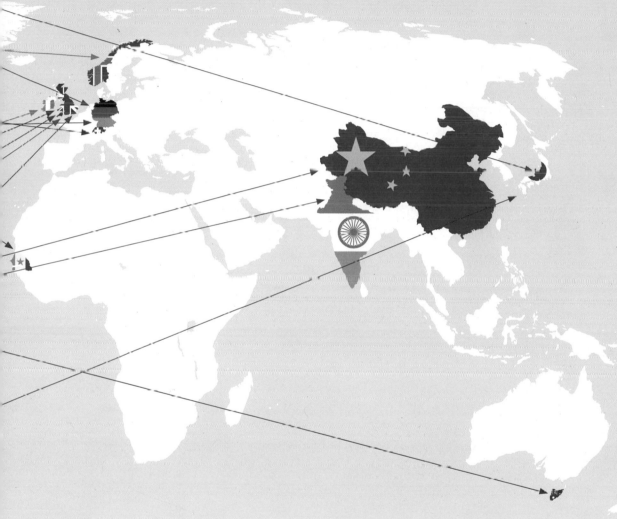

I have come to believe that we must take bold and unequivocal action: we must make the rescue of the environment the central organising principle for civilisation. Whether we realise it or not, we are now engaged in an epic battle to right the balance of our Earth, and the tide of this battle will turn only when the majority of people in the world become sufficiently aroused by a shared sense of urgent danger to join an all-out effort.

Former Vice President of the United States, Al Gore

Introduction:
The case for recycling

by Lucy Siegle

As a concept, recycling has it all. If you're a philosophical kind of consumer, for example, you'll know that 'new' products aren't really new. They are just matter and energy, shifted about and transformed from one state to the other. In which case, why not shift that same matter and energy into something else afterwards?

Recycling, then, fuses philosophy with logic. It also ticks just about every 'sustainability' box there is – it's a means of extending precious non-renewable resources, lessening the impact of consumerism on a beleaguered planet, and still gives us opportunities to forge ahead with new industries and new ideas. It breeds creativity and ingenuity and there's even magic: recycling as latter-day alchemy, turning rubbish into something special.

In fact – you might be thinking – what's not to like? Why aren't global recycling rates sky high, why aren't I dressed head to foot in pre-loved fibres and sipping my latte from a cup that was once a telephone directory and will soon be insulation?

Plato probably asked himself similar questions – although not the latte one. Around 400 BC, he was already on a sustainability trip. Not only did he write extensively about soil erosion and how goats and deforestation were destroying the arability of ancient Greek farmlands, but he also advocated the recycling of materials. Sadly landfill boasts an even more ancient pedigree. The first, in the Cretan capital of Knossos, opened in 3000 BC. The fact that landfill got there first might explain why Plato wasn't able to upgrade recycling into some sort of universal truth. That's too bad. Think how different things would have been if recycling had been woven into the fabric of civilisation. But you can just imagine how it happened. Plato would be busy expounding the concept of recycling waste down at The Academy, only to be shouted down by landfill enthusiasts. Same as it ever was.

> If you're a philosophical kind of consumer, you'll know that 'new' products aren't really new. They are just matter and energy, shifted about and transformed from one state to the other.

Humankind seems keenest on recycling when resource extraction is difficult. Anthropological studies of rubbish dumps in ancient Britain show tiny amounts of the stuff of life – ash, broken tools and pottery – suggesting that ancient societies recycled as much as they could.

This broken phone has been cleverly recycled into a fan. In developing countries, waste is often treated with more respect than in the West. (see Case study 7, p. 110)

Plato's enthusiasm for recycling does live on, albeit in quite a small way, in the form of Plato's Closet, a successful chain of American stores that have managed to make clothes recycling appeal to a young, fashion savvy market. Certainly an achievement, but probably not what Plato had in mind.

Despite all its plus points, the concept of recycling has been batted backwards and forwards for generations. Humankind seems keenest on recycling when resource extraction is difficult. Anthropological studies of rubbish dumps in ancient Britain show tiny amounts of the stuff of life – ash, broken tools and pottery – suggesting that ancient societies recycled as much as they could. There is also evidence of bronze scrap recovery systems in place from 2000 BC; and in the 1500s Spanish copper mines began using scrap iron for cementation of copper, a recycling practice that still survives. There is also evidence that composting – the ultimate recycling – was widely practised in China.

It was the Industrial Revolution that put the brakes on recycling in a big way. It marked a watershed, when materials became more ubiquitous than labour and suddenly there was no need to worry. You could also argue that it sowed the seeds for the global economy which continues to have an impact on waste today, as global trading went up several notches. When the world, and its associated mineral deposits and natural resources, is your oyster, why worry about eking out resources?

But now we do need to worry. Over the past 50 years humankind has altered ecosystems more extensively than at any comparable period of time in history. This has largely been in order to meet growing demands for food, fresh water, timber, fibre and fuel, resulting in a substantial and largely irreversible loss in the diversity of life on earth. If you're looking for further evidence, you can always read the 2005 *Millennium Ecosystem Assessment*, produced by 1,300 researchers from 95 countries. As this volume weighs in at some 2,500 pages, it's probably best to turn to the executive summary, the gist of which is that two thirds of the earth's life-sustaining ecosystems are under threat of collapse. If this is not a driver for recycling, I don't know what is.

The days of abundant, cheap raw materials which spearheaded the Industrial Revolution are over. For environmental thinkers, such as Amory Lovins from the Rocky Mountain Institute, the next revolution is one of new materials, of designing from cradle to cradle, as opposed to cradle to grave – i.e. landfill. The problem with conventional capitalism, as far as Lovins is concerned, is that it is a system which avoids the protocols laid down by nature, such as recycling, and therefore precisely the system which imposes limits on increased prosperity. Switch to a system which recognises natural capitalism – i.e. the importance of natural resources which are relentlessly recycled through the system – and those limits are lifted. In this context, recycling represents a kind of ecological 'get out of jail free card' and one that needs to be used quickly.

A tractor ploughs through landfill.

As a species, human beings are possibly terminally unique. We are for example the only animals that mine for resources, unlike other species who produce what they need, as spiders produce their webs, or utilise what's around.

Because materials collected for recycling have already been refined and processed once, this means that manufacturing the second time around is usually cleaner and less energy-intensive.

Then there's the fact that resources will not have to be extracted again. Increasingly, this last point will prove to be crucial, both in terms of extending and preserving dwindling resources and protecting the last remaining pristine habitats.

As a species, human beings are possibly terminally unique. We are, for example, the only animals that mine for resources, unlike other species who produce what they need, as spiders produce their webs, or utilise what's around. Unless we learn to recycle resources, the danger is that mining operations will wade even further into crucial habitats. This can be seen already in the oil industry – the US administration dreams of annually extracting oil from the Arctic National Wild Life Refuge. Meanwhile, the mining giant Rio Tinto plans to push on into areas of south east Madagascar to mine titanium (used for whitening paint, toothpaste and paper), putting 200,000 species of flora and fauna, unique to this part of the planet, at risk. The island is one of the world's three 'hotspots' for biodiversity, and yet we may destroy it to get at resources to make products used in our everyday lifestyles.

Put starkly, resource mining and stripping is an environmentally destructive practice which ultimately threatens the very survival of our species and the planet. So why aren't we recycling like fury? What the recycling debate has traditionally lacked, among the maelstrom of opinion and theory, is a clear idea as to exactly how recycling can work and what it can achieve. Precisely, in fact, what this book offers, delving deep down to uncover the real potential.

A Greenpeace protester paints graffiti on an incineration facility in Sheffield.

There's no denying that green and eco-based issues are currently very hot. They mirror our global angst about climate change and resource depletion.

There's no denying that green and eco-based issues are currently very hot. They mirror our global angst about climate change and resource depletion. I believe that most people are interested in solutions, albeit ones that don't interrupt daily life too much or require wild sacrifices or the wearing of tie-dye. Just five years ago I kept my tree-hugging propensities to myself, now I'm 'out', writing about environmental and ethical living in the national press. Recycling, and the cohort of issues that accompany it, remains one of the readers' favourite topics. How can they recycle more, and more effectively? How can they be sure that the local authority is truly recycling their carefully washed out jars and separated paper? And how can we lobby our local authority to do more?

But overall, when it comes to suspicion and conspiracy theories, recycling rivals the assassination of John F Kennedy. I went to a party recently where the subject strangely drifted on to recycling (another thing that wouldn't have happened five years ago). A contemporary of mine explained conspiratorially that a friend of a friend had seen the distinctive orange recycling bags we dutifully fill in our area of London every week, lying cheek by jowl with the normal black bags in a landfill, just a few miles outside of London. These are the apocryphal tales of phoney kerbside collections as a means of boosting a local authority's recycling collection rate, which spread suspicion.

But it's hardly surprising that there's scepticism. Few of us know what happens to our recycling — we all know what spewing, toxic, landfill dumps look like because we've seen the pictures, but few of us have seen inside a state of the art recycling plant. In this book we'll shed light on that too.

For residents of Britain and Ireland, there is a sense that we have spent so long languishing somewhere near the bottom of European recycling leagues (although Ireland has steamed up the charts recently largely courtesy of its tax on plastic bags. See Case study 9, p. 136) that the task of reversing such trends can seem insurmountable. In fact, 2004 recycling rates showed a rise from 3.2 million tons to 3.7 million tons of household rubbish recycled, but the UK will not be able to shake its 'dirty man of Europe' soubriquet overnight, especially as German households still manage to recycle four times as much of their domestic waste.

Municipal waste management in the European Union

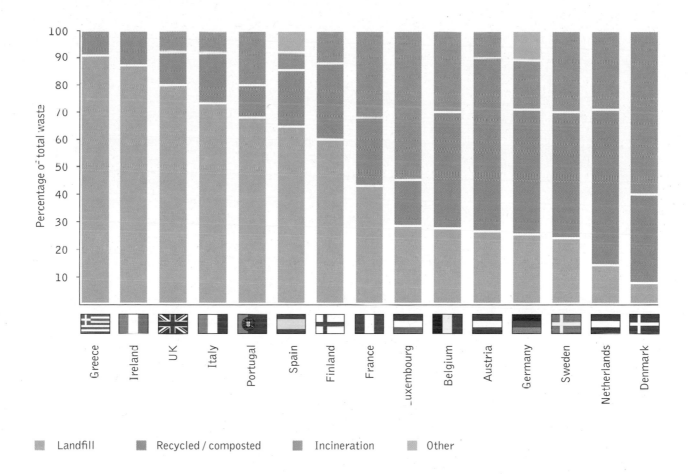

Landfill   Recycled / composted   Incineration   Other

The UK will not be able to shake its 'dirty man of Europe' soubriquet overnight, especially as German households still manage to recycle four times as much of their domestic waste.

Britain, Portugal and Greece sit at the very bottom of the league tables of municipal waste management.

In the UK, the first bottle banks opened in the 1970s, but the systems that followed were largely piecemeal and regionally specific, as each local authority introduced a cost-determined version of recycling, and some failed to introduce anything at all. Lately, things have improved with 111 authorities in England (31 per cent of the total number) offering a collection of two or more materials to at least 95 per cent of the households in their area. These are the local authorities that meet the 'Ruddock Test' which will make these levels a statutory minimum by 2010.

If we were writing a report card, it would say something along the lines of "shows some capability but MUST do better".

In an age when the city of Canberra and parts of New Zealand have achieved nearly total recycling participation and are now focused on the holy rubbish grail of 'zero waste', these recycling rates might be an improvement, but they're very far from spectacular.

At the time of writing, no authority in Wales meets the Ruddock Test standards for example.

If we were writing a report card, it would say something along the lines of "shows some capability but MUST do better", particularly if you take into account that other areas of Europe do substantially better. Overall, Europe produces two billion tons of domestic rubbish every year. Holland reduced its per person production of post-consumer waste from 497 kilograms in 1990 to 390 kilograms in 1993, and is pursuing the goal of 75 per cent waste reduction and recycling. Sweden is aiming for 70 per cent reduction by 2005.

What I hear people ask is, "is our rubbish really sorted? Or is it just exported?". In common with food, apparel and just about every other modern day lifestyle-enabling product or service, waste is a global issue. It's not yet a commodity – although in some countries it is revered far more than others; in Brazil for example, it is said that an aluminium can once thrown away never touches the ground, as it will be picked up by one of the thousands of *catadores* who make their living sorting trash for recycling. But the practice of shipping waste to poorer countries for a form of recycling, in effect just exporting the problem, is unacceptable to most people in the UK who collect paper and wash out jars in good faith.

The landmark recycling schemes are the schemes that keep it local. These are the ones that offer proof of a tight closed loop system, and generally the antithesis of the 'rubbish-for-profit' type operations instigated by the huge waste contractors, who control the lion's share of the UK's recycling operations today. Wye Cycle, the brainchild of recycling enthusiast Richard Boden, is one of the small, successful schemes. Operating across 2,000 households in Surrey in the south of England, it has introduced separate kerbside collections of five materials plus an electronics collection point. Through a system of composting and repairing redundant items and selling them back to the community, the scheme has got each household's landfilled waste down to less than 260 kilograms per year, as opposed to the national average of one ton. Another

Household waste and recycling: 1983/4 and 2003/4. United Kingdom

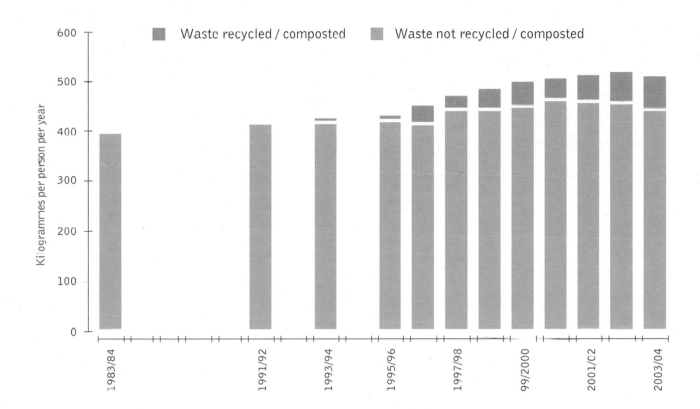

For recycling to keep moving forward, we need an enlightened mindset, one where recyclers and designers think of recycled materials as a true resource, and one that will add value to products rather than detract from them.

example is environmental charity Bioregional, which has pioneered an office paper collection, 'The Laundry', which will recycle office paper through small local mills (or mini mills) rather than the giant globalised corporations that now control the paper industry, and sell this product back to local firms. It is an attempt to buck the national trend of low paper recycling rates. Despite the fact that office paper is a prime candidate for recycling – being high grade it can successfully be recycled down the chain five or six times, 86 per cent is sent directly to landfill to be mummified.

Successful recycling schemes don't operate in a vacuum. They foster a complete, and creative, approach in terms of finding markets to sell recycled waste into and products that can be sold back to the community. For recycling to keep moving forward, we need an enlightened mindset, one where recyclers and designers think of recycled materials as a true resource, and one that will add value to products rather than detract from them. Breakthroughs can appear to be small – a manufacturer able to incorporate ten per cent instead of just five per cent of recycled rubber from scrap tyres

With the exception of glass, which can be recycled into perfectly good glass, many materials work best recycled into a different incarnation.

could bring millions more scrap tyres into the materials loop every year. But above all, recycling should give designers a chance to showcase their ingenuity. With the exception of glass, which can be recycled into perfectly good glass, many materials work best recycled into a different incarnation. Newspapers, for example, make great cereal boxes or insulation materials. To close the loop, consumers need to be attracted to recycled products to buy them, and keep the market moving forward. By now, most of us are probably familiar with the frequently quoted fact that it takes 25 plastic bottles to make a medium sized fleece jacket. This is great. But once we have our fleece we're unlikely to need another one. We need more recycled products, greater creativity and variety.

Newspapers can be recycled into lower grade paper such as egg cartons, whilst office paper can be recycled back into office paper. Plastic PET bottles can be recycled back into bottles or reincarnated as fleece.

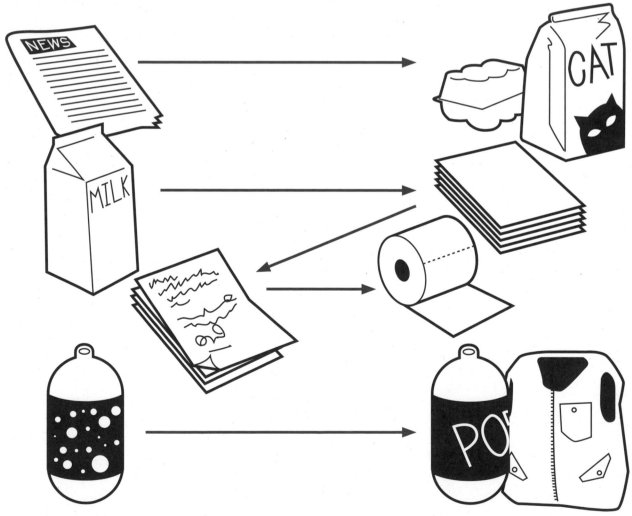

Naturally, this fledgling market requires continuous, assured levels of supplies. It needs us to recycle – which begs the question, what kind of recycler are you? Are you committed, revolutionary and rebellious, or a fair-weather recycler, one who is happy to sort your trash as long as everything is straightforward and you don't have to rinse out baked bean cans, or tax your brain, working out if a piece of hybrid packaging qualifies as paper or plastic?

What it doesn't need, as many New Yorkers will tell you, is inconsistency. In the 1980s through to the early 90s, recycling had many shiny eyed proponents, especially in official departments. Arguably they saw recycling as a potential vote winner, and certainly as a panacea to the ever-increasing mounds of domestic rubbish generated in urban homes. Recycling was firmly wrapped in a bullet-proof vest, immune to criticism. Then things started to change as increasingly the benefits, and particularly the cost of recycling, which is muddied by complex economic debates and relative values, were called into question. The newspaper headlines of 1990 announced that "Garbage is Gold" when just six years later, the *New York Times Magazine* ran a seminal anti-recycling piece entitled, "Recycling is Garbage". Things weren't moving fast enough or cheaply enough.

The most stark example of an official change of heart occurred in 2002 when New York City mayor, Michael Bloomberg, announced plans to axe glass and plastic recycling to save money. According to a report by the city's Independent Budget Office, New Yorkers were spending up to $48 (£27) per ton more to recycle than it would cost to landfill or incinerate. Lay the environmental and health gains to one side – as so often happens – and the fiscal economics just didn't add up.

What happened next could either amuse or infuriate, depending on how you view it. In March 2004, Bloomberg announced that New York's glass recycling programme would resume alongside a mixed recycling programme including plastics, metal and paper. In effect he decided to give back the system that he had taken away. The recycling-free hiatus, officials argued, had given waste contractors and vendors dealing in recyclables time to get their own houses in order, and to become more efficient and cost-effective.

What kind of recycler are you? Are you committed, revolutionary and rebellious, or a fair-weather recycler, one who is happy to sort your trash as long as everything is straightforward?

New York's recycling programme lost impetus during the gap in programmes, and hundreds of thousands of former recyclers regressed to dumping their trash unsorted.

There's some truth in this, because recyclables are now collected at $51 (£29) per ton, which is half the previous cost, but recycling proponents deeply annoyed by the removal and replacement of the scheme, allege this is more to do with an upturn in market prices for glass, plastics and paper. Furthermore, they claim, New York's recycling programme lost impetus during the gap in programmes, and hundreds of thousands of former recyclers regressed to dumping their trash unsorted.

US Trends in Waste Generation, Recovery and Disposal

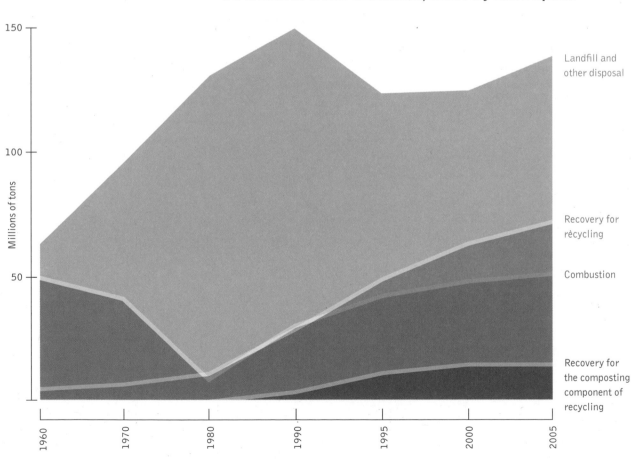

Millions of tons

Landfill and other disposal

Recovery for rècycling

Combustion

Recovery for the composting component of recycling

These are behaviours, it can be convincingly argued, that are extremely difficult to unlearn. Certainly, other US cities seem to have learned from New York's mistake. When Cincinnati officials mooted an idea of dispensing with kerbside recycling as part of a cost cutting programme they faced public outcry. Instead, the officials cut elsewhere, and actually diverted money into recycling.

Other US states that seemed previously disinterested have actually become stalwarts of municipal recycling. Kansas, for example, was particularly slow on the recycling uptake, but now only collects two conventional rubbish bags per household per week, as part of a 'pay-as-you-throw' scheme. And the economic incentives to recycle appear to be getting stronger too. A recent US study, the St Louis Recycling Economic Information Study, found that recycling in the St Louis metropolitan area provided nearly 16,000 jobs and generated almost $5 billion (£2.8 billion) a year in revenues.

This type of economic boom is not normally associated with recycling, but studies show that recycling can be made to pay, especially if the costs of landfill are raised to reflect their hazardous environmental impact – something which happens increasingly in Europe. Another US study of Baltimore, Washington DC and Richmond showed that each ton of landfilled waste generated $40 (£22) in tipping fees and just 13 jobs per 100,000 tons. If this waste was recycled it generated $120 (£69) per ton in revenue and 79 jobs per 100,000 tons. So in fact it's hardly surprising that increasingly legislation finds in favour of recycling.

Reports, then, on the demise of recycling have proved to be greatly exaggerated. As waste levels continue to increase against an alarming environmental backdrop, recycling has once again stepped forward as a kind of ecological caped crusader.

Perhaps finally this is the chance for recycling to get us out of the gargantuan hole we would normally try to fill with our trash.

Perhaps finally this is the chance for recycling to get us out of the gargantuan hole we would normally try to fill with our trash.

Barriers to participation do exist – again, this book will hopefully help you to remove any that continue to block your way to becoming a committed recycler. At The UK's Centre for Sustainable Consumption at Sheffield Hallam University, Janet Shipton has researched some of the barriers to reusing and recycling objects in our home. She uncovered the type of 'twilight' zones that I know I'm familiar with. These are the areas – typically under the kitchen sink and hallway cupboards – where we save packaging, usually plastic bags and bottles. This is the waste we know we'd like to recycle, but we're not always sure how to, or perhaps we run out of steam and so it gradually accumulates in these 'resting' places. Further evidence that the road to recycling is paved with good intentions.

Without packaging of course, there would be no contemporary recycling story. There is no doubt that packaging has revolutionised industries, most notably the global food industry – it's not beyond comprehension for a top chef in London to pick his fish via satellite from a fish market in south east Asia and have it packed in ice, bundled up in polystyrene and air freighted. As consumers, we routinely buy a four pack of kiwi fruit from

The typical contents of a UK bin; packaging is equal in proportion only to kitchen waste.

Packaging 40%

Paper 20%

Putrescible (kitchen waste) 40%

Packaging, particularly plastic wrap, is after all the skin of commerce. Without the boxes, bags, trays and vacuum packs, filled with modified atmospheric gases to enable longer shelf life, where would we be?

New Zealand, bundled up in cellophane on a wax lined tray, from the local supermarket. The packaging facilitates this global food economy and in a sense obscures the truth of the situation; that each kiwi fruit has emitted five times its own bodyweight in carbon dioxide emissions through its epic journey, has consequently exacerbated climate change and ties us in ever tighter to a global oil-reliant economy. Speaking from an environmental and ethical perspective, none of these features are attractive or sustainable.

But criticise packaging and you run the risk of being seen as some kind of Luddite. Packaging, particularly plastic wrap, is after all the skin of commerce. Without the boxes, bags, trays and vacuum packs, filled with modified atmospheric gases to enable longer shelf life, where would we be? And don't answer "eating local, seasonal, sustainable produce provided with half as many toxic chemicals" as that might be considered facetious.

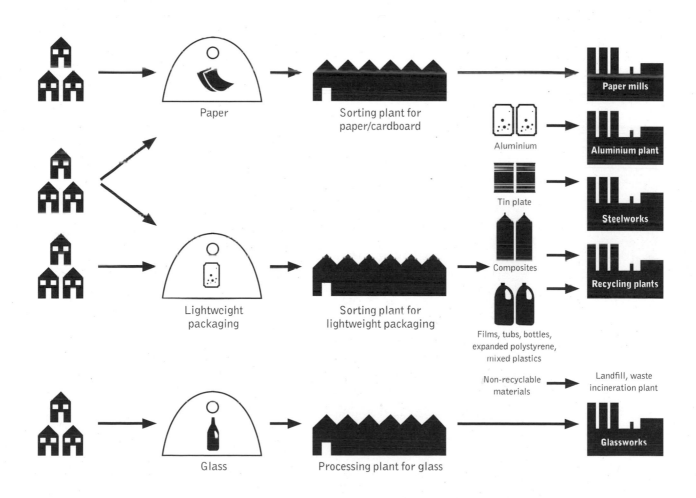

Paper

Sorting plant for paper/cardboard

Paper mills

Aluminium

Aluminium plant

Tin plate

Steelworks

Composites

Recycling plants

Lightweight packaging

Sorting plant for lightweight packaging

Films, tubs, bottles, expanded polystyrene, mixed plastics

Non-recyclable materials

Landfill, waste incineration plant

Glass

Processing plant for glass

Glassworks

For better or for worse, packaging is embedded in our lives, something we could forgive if the food processing and packaging industries wholeheartedly addressed recyclability. Your average piece of packaging fulfils a whole load of criteria. It preserves the product hygienically, allows it to be transported and stored, conveys a marketing message, deters shoplifters and should also promote recyclability. However, increasingly, the recyclability goal falls to the bottom of the list, or falls off completely. Increasingly, packaging prioritises functions other than sustainability. They represent 'smart' packaging. This might, for example, take the form of a razor packet embedded with radio frequency identification technology to deter shoplifters. Alternatively a packet might use commingled plastics fused together to make a fun shape because its primary function is to shout desirability to the under fives and harness the effect of so-called 'pester power'. In either case, the materials inherent in the packaging will be too complex to be recovered. Add to this the type of single-use convenience packets that are the complete antithesis to sustainable recyclable packaging, and you begin to wonder where it's all going wrong. They are part of the one night stand convenience culture in which, US writer and environmentalist Wendell Berry warns, "the histories of all products will be lost. The degradation of products and places, producers and consumers is inevitable." So while UK schools spend £39 million ($70 million) picking up litter from playgrounds, the processed food industry giants are gunning for the 'lunchables' market, considered to be the next big thing. It's out with the good old fashioned lunchbox, that stalwart of sustainability which could be reused for an entire school career, and in with single-use disposable, commingled plastic containers. How does recycling fit into this type of trend? Sadly it doesn't.

The onus then is on consumers, to stand up to unsustainable packaging trends. Living more sustainably means recycling more and different materials. Like the old weight-loss adage, 'a moment on the lips, a lifetime on the hips', it's about realising that whilst a coffee might take five minutes to drink, the cup will take hundreds of years to biodegrade. Sometimes it's about altering behaviours, and sometimes about questioning our habits. For example, it takes ten times more energy to make a ton of textiles than a ton of glass, and yet few of us think about recycling clothes. Increasingly, creatively-minded businesses, many not-for-profit or charities, are stepping in to fill the breach. Textiles for

Add to this the type of single-use convenience packets that are the complete antithesis to sustainable recyclable packaging, and you begin to wonder where it's all going wrong.

TRAID clothes banks can be found on street corners all around the UK. (See Case study 12, p. 166)

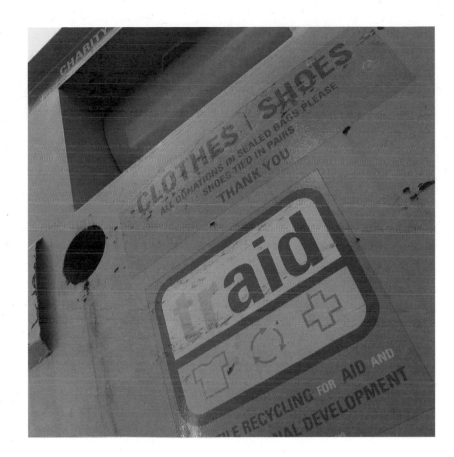

Aid and International Development (TRAID) is a case in point, not only collecting clothes through its TRAID clothes banks, but also adding value to the recycled products which are altered and customised by designers under the TRAID Remade label.

Increasingly, recycling efforts will be driven by legislation as landfill tax puts pressure on businesses to create less waste in the first place, and they find something more constructive to do with the waste they do create. The construction industry, one of the biggest consumers of resources and producers of carbon dioxide emissions, is notably in the firing line. It will be up to projects such as the UK's National Community Recycling Project to step in and fill the breach. It trades in scrap wood from construction projects, selling it on to domestic consumers for their own homes, saving hundreds of thousands of tons of wood from landfill and saving thousands of trees in the process.

It's always worth emphasising that one of recycling's main assets is precisely that it is not landfill. Wherever there is large-scale resource extraction, landfill has followed. In the

UK, we're used to gaping holes in the landscape, they're a legacy of decades of coal mining. As soon as a mine had been liberated of all its deposits, it became an opportunity; the perfect conduit for our rubbish. Landfills also appear on the sidelines of heavy industry. When times are good, they are part of the infrastructure; during periods of economic decline they make these areas appear little more than great big rubbish dumps, where inevitably the poorest members of the community are also housed. In the US, Michigan is a good example, providing a receptacle not just for waste from the immediate areas, but for Toronto, across the Canadian border where around 1.5 million tons of garbage makes its way to Michigan landfill each year.

One thing we generally don't have a problem grasping is the visual significance of landfill dumps. They are testament to the fact that our waste problem is out of control and not going anywhere. In fact there are only two human made creations visible from outer space, one the Great Wall of China, the other Fresh Kills, the now defunct landfill, primarily composed of New York's household waste which according to many was an illegal dump, situated in ecologically significant wetlands in the first place.

There are plans now to rehabilitate Fresh Kills as a nature reserve, grassing over the mounds of household rubbish. In fact this is not uncommon; in 2002 senior waste management figures estimated that at least three million tons of illegally dumped materials, primarily from farms and businesses, had been used to mould the contours of a number of the UK's brand spanking new golf courses, shopping malls and housing estates. Talk about burying the problem.

This kind of greening over of massive heaps of waste also begs the question: what lies beneath our feet? – and brings us back to a few unpalatable truths about landfill. Waste doesn't go away. In a landfill setting it's estimated that fruit and vegetables take up to two years to decompose, plastic bags (of the very thin kind) take around 20 years; paper racks up 50 years; for aluminium cans and foil add around 100 years – and plastic bottles, one of the biggest contemporary waste pests, can take up to 450 years.

If only biodegradability was the only problem. It would be inconvenient to have the nation's fridges and apple cores hanging around, but not insurmountable. Of course, that's far from the

A new campaign entitled Recycle Now has been launched in the UK by WRAP (Waste and Resources Action Programme). This is a detail of one of its posters promoting can recycling in Britain.

A projected view of Fresh Kills landfill on Staten Island, once it has been regenerated.

only issue. In landfill our cast-offs, from organic matter and empty batteries to the most innocent looking textiles, become toxic. Leachates, potentially toxic liquids from organic waste, seep out of landfill sites and pollute precious groundwater sources, eventually finding their way into our primary watercourses.

Meanwhile the non-organic waste – the cushions, hairdryer parts, batteries, holey socks, discarded toothbrushes, cotton wool, Teflon coated school skirts and tin cans – whoop it up in a bizarre and unnerving mix of potentially toxic chemicals. When you consider that the American Chemical Society has recently registered the 10 millionth man-made chemical and that the Chemical Industries Association estimates the average British family buys around £25 ($45) worth of chemicals every week, you can see just how prevalent chemicals are in our household rubbish. What happens to these chemicals in this gigantic, fermenting cocktail? Nobody really knows; suffice to say that modern life is all about rubbish. An average American generates about two kilograms of waste per day, and will throw away 600 times their adult weight in garbage over their lifetime, or 40,000 kilograms of rubbish. In terms of

Greenpeace protesters target an incinerator in Britain, calling attention to its toxicity.

At frequent points in history, incineration has reared its rather ugly, polluting head as the way of dealing with growing rubbish mounds.

resource use, between 1940 and 1976, the US consumed more minerals than the whole of humanity had managed to get through in its entire existence. Each person in the UK throws away 4.5 times their own body weight in rubbish each year. This would be an alarming enough waste pandemic on its own, but there's a further catch. To get a truly accurate picture of the average waste footprint, you have to consider the knock-on effects of your rubbish. Add in the waste created by transportation, subsidiary packaging, and all the little chains along the way, such as packaging and marketing initiatives and the rubbish generated by an average British adult each year weighs in at near 200 times the average body weight, or 20 tons per person. And it's growing by 3.2 per cent a year.

At frequent points in history, incineration has reared its rather ugly, polluting head as the way of dealing with growing rubbish mounds. It is still advocated in some countries, particularly in France where local incineration is popular, as proponents point to the way it reduces the volume of waste by up to 90 per cent. However, this doesn't remove the problem of the toxic ash that's left behind which still has to be landfilled. Nor the fact that a huge body of research connects incineration with severe health problems caused by toxic fumes and poisonous ash. The increasing amount of complex plastics in the waste stream doesn't help – when burned they produce dioxins, one of the most pernicious off-gases linked to cancer and infertility.

Recycling not only has a lot of pluses, but it's also the logical contender to the waste throne. And arguably now is when we need it most, in our hour of ecological need if you like. Naturally there are lots of other schemes vying for attention. The mainstream packaging industry for example, along with major retailers, likes to promote the idea of biodegradable packaging. Which seems like a good idea on the surface but does it really stand up to the test? This would still presumably involve landfilling plastics, biodegradable or otherwise, and unfortunately landfill doesn't offer the right conditions for even cornstarch plastic to biodegrade.

What's needed is a scheme that's transformative, revolutionary, in alliance with conservation, and makes a virtue out of trash. Recycling in fact! In most areas of the developed world there has been a slew of initiatives to rebrand recycling as something cool, and catapult it into the national psyche. Popular eco-aware brands such as Innocent Smoothies come in recyclable Tetrapaks to get the message across to young customers with a logo reading, "Dudes love dustbins (especially recycling ones)". This is all laudable stuff, but shouldn't detract from recent studies showing that people will recycle more if the facilities and processes are available and clear to follow. That, after all, is the key.

And arguably, even more important than finding recycling poster boys and girls is to make the market for recycled goods aspirational, interesting and accessible.

As consumers it's important that we can bulk buy recycled toilet roll — after all, the UK alone gets through 13 million rolls a week — but that we can buy aspirational products too. Traditionally recycling involves 'downcycling' — that is, with each process the original resource, be it oil based plastic or office paper, loses value. As this is of limited appeal to manufacturers, waste specialists, regulators and consumers, for recycling to progress into the mainstream we need more people to take an opposing view: i.e. 'upcycling', when the design process recycles upstream in order to add value to the material. This is an idea that Dutch design consortium, Eternally Yours, works to promote and one favoured by the inspirational US eco thinker and green architect William McDonough, who together with industrial chemist Michael Braungart, argues against the manufacture of recycled products where the eco message is more important than the usefulness or the value of the product. In other words, the agenda to recycle should not supersede the rest of the design process. It is not enough for an object or material to *seem* ecologically benign.

Recycling has lots of lessons to learn. Some of them will be of the tough love variety, through legislation, as in the case of Seattle. From 2006 if you leave out contaminated recycling bags containing the wrong kind of rubbish (i.e. non-recyclable), expect the bags to stay there and to be reported to the council, following the example of schemes already operating in Switzerland and parts of Austria.

But some of the lessons we learn should be from nature, allying recycling to biomimicry, a relatively newly defined branch of ecology that concedes that nature, with 3.8 billion years of

And arguably, even more important than finding recycling poster boys and girls is to make the market for recycled goods aspirational, interesting and accessible.

practice, has a ready-made syllabus for ensuring our survival. The core idea is simple. If you want to build the perfect solar cell, analyse and copy the structure of a leaf, the ultimate solar appliance. Similarly if you want to know what to do with waste, look to nature where there's no such thing. One creature's cast-off becomes another's home or food supply. Biomimicry, creativity and contemporary relevance – recycling has it all.

'Sunday Papers' chair, designed by David Stovell, illustrates innovative ways of reusing materials in design. (See p. 218 for details)

RECYCLE

NEWSPAPER

FOR INFORMATION 566-1503 epa

Thank God men cannot as yet fly and lay waste
the sky as well as the earth!

Henry David Thoreau, 1817–1862

plastic
bottles

Materials and Case studies

telephone
directories

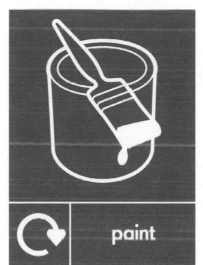

paint

food &
drink cans

clothes

newspapers
& magazines

garden
waste

Paper

What is it?

The word paper comes from papyrus, the ancient Egyptian writing material woven from papyrus plants. Forms of paper were known to exist as early as 3000 BC, and were made of materials ranging from sheepskin parchment, to silk, to hemp rags. Throughout the centuries paper was considered a luxury item, until the invention of a steam driven paper making machine in the nineteenth century, which could mass produce paper using fibres from wood pulp. This was known as the Fourdrinier paper making machine, and it remains the basic model for all modern paper production.

The Fourdrinier paper making machine at the turn of the century and today – modifications have been fairly minimal.
Previous pages: graphic icons from the Recycle Now campaign in the UK.

Why recycle?

Paper today comes in all colours and consistencies. The general categories are:

- Computer print-out
- White office paper
- Cardboard
- Newspapers
- Coloured paper
- Magazines and pamphlets
- Kraft waste (brown paper bags and wrapping)
- Paper sacks

Most virgin (un-recycled) paper is made from wood grown in 'sustainable forests'. The problem with using virgin paper is not that trees are being senselessly cut down, but the ways in which these sustainable forests are grown and maintained.

Many diverse, old-growth natural forests throughout the United States, Canada and the developing world are being cleared in order to be replaced by mono-culture plantations, usually of conifers, that employ toxic chemical herbicides and fertilisers to maximise growth. Roads need to be carved through the natural forests to allow logging routes into the industrial harvests. The end result is the destruction of wildlife habitats and ecosystems. In the UK, for example, paper is sourced in Scandinavia. Only five per cent of Scandinavian old-growth forests now remain, and even these continue to be logged.

Only five per cent of Scandinavian old-growth forests now remain.

The problems with virgin paper do not stop there. In order to bleach the paper, powerful chlorine based chemicals are used, emitting vast quantities of toxic pollutants called dioxins into the atmosphere. Even after the paper has been thrown away it continues to do damage, rotting in landfills and creating methane gases which contribute to the greenhouse effect.

Paper usage today has sky-rocketed. Industrialised nations, comprising 20 per cent of the world population, use close to 90 per cent of the world's printing and writing papers. After garden and kitchen waste, paper and cardboard comprise around 18 per cent of rubbish in most bins, amounting to approximately four kilograms of waste paper per household, per week in the UK – and it's on the rise.

Global production in the pulp, paper and publishing sector is expected to increase by

77% from 1995 to 2020.

It is essential to move paper production and consumption towards ecologically responsible and sustainable processes.

But does recycling really make a difference?

There is some debate surrounding the usefulness of paper recycling. Critics suggest that producing recycled paper uses more energy than virgin paper production, is more polluting and may make a greater contribution to climate change.

A landfill stretches into the horizon – as yet recycling is the only real alternative to landfill.

The evidence however, points strongly in the other direction, estimating a saving of anywhere between 28 to 70 per cent of energy consumption. This is mainly due to the energy that is required by the pulping process, turning wood into paper.

Although no conclusive research has been done on the pollutants emitted in the recycling process, here too the estimates come down heavily on the side of recycling, pointing out that the air emissions are cut down to virtually nothing, and the water emissions (effluents) are significantly reduced as the recycling process becomes more sophisticated. Recycled paper is usually not re-bleached and this also reduces the amount of dioxins released into the environment.

Another argument the critics put forward, is that recycling doesn't make sense financially, and can't compete with new products. There is some truth in this, as recycling is more expensive than ordinary trash disposal, but a long term view indicates that as technologies improve and more people become involved, the cost of recycling will drop, and as space becomes more of an issue, landfill will become more expensive. This is an investment in the future, in all respects.

It is estimated that for every ton of paper recycled you are saving:

3000 litres of water

3000 — 4000 KWh electricity

95% of emissions

2.5 cubic metres of landfill space

17 pulpwood trees

So the answer is yes!

And the more we recycle the more efficient and productive the process will become.

The recycling process for paper

Once your used paper arrives at the recycling plant it is sorted and graded according to its quality.

These classifications are based on weight, colour, usage (industrial, domestic, food packaging etc.), raw material (whether the paper has already been recycled once or whether it's made of virgin pulp), surface treatment (coated/uncoated) and finish (fine/coarse).

Paper products such as corrugated cardboard, colour magazines and telephone books are all low grade products.

Newspaper and certain types of office paper are medium grade.

Wood-free white paper is high grade.

1
After the sorting, the paper is then pulped using water, chemicals and heat, until it breaks down into fibres known as cellulose.

2
The pulp is cleaned of all contaminates, such as glue and staples.

The final product varies depending on the grade of paper originally used. Office paper may be directly reincarnated as office paper. Telephone books and certain cardboards often become egg cartons, insulation or animal bedding.

About a third of the original material placed in the pulper will end up as sludge. This is the solid waste comprising small fibres, the bits of glue, staples and the ink that have been washed off during the recycling process. Traditionally this waste has been consigned to the landfill. However, new alternatives are currently being explored, including composting, incineration, and recycling in the form of gravel and concrete.

10
The paper is then ironed flat into large sheets, rolled and ready to be used.

9
Felt covered rollers squeeze out any remaining water.

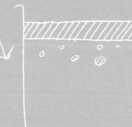

3
It is filtered through a
screen, spun at high speed
in cone shaped cylinders.

4
Finally, it is de-inked.
There are two ways of de-
inking paper; it is either
washed, using chemicals
which separate the ink
from the paper; or a
technique called floatation
is employed, in which air is
passed through the pulp,
producing a scummy foam
in which the ink collects,
and can be skimmed off.

5
If the end product is
destined to be white
paper, a bleaching
process will occur in
which chlorine is used.

6
At this point, a small
percentage of virgin pulp
may be added to the mix.
The reason for this is that
the fibres that make up the
paper are shortened with
every pulping process,
and the quality is lowered.
The virgin pulp adds
strength and smoothness
to the material.

7
A large amount of
water is added, and the
mixture is sprayed onto
a fast moving screen.

8
As the water drains,
the fibres begin to
bond together.

What's the future of paper recycling?

Although major advances have occurred in paper recycling, at the moment only about a quarter of paper in Britain is being recycled. In America this figure is even smaller.

The laws governing paper recycling vary enormously from country to country. In Britain there are no laws governing recycling — only targets for municipal waste as a whole. In certain states, such as Hawaii, it is compulsory for offices to recycle their paper, and in rare cases, such as that of Wisconsin, all paper has been banned from landfills. These laws, as well as people's willingness to participate in recycling schemes, will have a direct impact on the way the paper recycling industries evolve.

'Shredded Paper Recyling Bin' designed by Lynn Kingelin. (See p. 217 for details)

The main problems facing paper recycling are:

- Lack of markets for the recycled product.

- Lack of funding for recycling.

- Lack of enthusiasm and participation from local populations.

- Lack of appropriate recycling facilities and sites on which to construct them.

Possible legislative changes could include putting a tax on virgin pulp, making it compulsory for local authorities to run comprehensive recycling schemes and ensuring that companies use certain percentages of recycled products.

There also needs to be investment in new technologies to improve efficiency and eco-friendliness of the recycling processes. Both in Japan and the United States, 'decopier' technologies are being explored. These will strip toner directly off paper (without any of the recycling process) so that a single sheet can be reused up to ten times.

Finally, each and every one of us needs to start being pro-active and taking responsibility for our own waste.

The United States recycles
36% of its paper

Japan recycles
50.3% of its paper

What can I do?

Firstly, clarify what local recycling schemes are available to you.

Many places will offer kerbside recycling schemes. In other areas, you may have to do your own transporting to a local recycling depot. Find out whether your depot requires you to sort the paper grades yourself. If they do, you will probably have to divide your paper waste into the following categories:

- White office paper
- Newspaper, magazines, telephone directories, leaflets
- Cardboard
- Mixed or coloured paper

Try and make sure that all staples, post-its, plastics and foils have been removed before you take it in. There are relatively few recycling plants with facilities for recycling juice or milk cartons, so check before you put these in!

Try not to use as much in the first place!
Reuse paper as much as possible:

- Reuse envelopes.
- Print on both sides.
- Cut down on unnecessary extras like fax cover sheets.
- Think twice before you print something out from your computer – do you really need a hard copy or can you read it just as well on screen?
- Collect scrap paper for notepads.
- If you are receiving a lot of junk mail, contact your mailing preference service to have your name removed from mailing lists.
- Adjust margins to print more on a page.
- When you do buy paper products, try to buy recycled.

The future of recycling depends on there being a market to supply. If you cannot find the product you're looking for, try ordering from the internet (see the directory section for hints and tips on where to go).

In the office

Many millions of tons of printing and writing papers are lost every year through offices, so implement an office recycling scheme:

- Think about best ways of paper collection in your office. Where can you put recycling bins, so that people make the most of them?
- Provide clear lists of what is and isn't recyclable (clarify what these are with your recycling service provider).
- There are often local waste paper merchants and national collection companies that deal with large quantities of paper. Contact whoever deals with your normal waste; many provide recycling services, and if they don't, they should be able to put you in touch with someone who does.
- If you have difficulties arranging a recycling service provider, contact other businesses in your area and try to set up a recycling cooperative.

Get your kids involved in recycling: Make your own recycled paper.

To make your own recycled paper you will need:

- Scrap paper or newspaper, shredded into five centimetre squares
- A food processor or a whisk
- Water
- A small plastic tub filled with ten centimetres of water
- A measuring cup
- White glue
- Food colouring or onion skins
- A piece of wire window screen or a coat hanger and stockings
- An iron

1 If you do not have a window screen, make your coat hanger into a rectangular frame. Take one leg of the stockings and stretch it carefully over the hanger frame. Make sure it's tight and flat. You'll need a frame for every piece of paper you make, so you might want to make a few.

2 Soak the paper in a bucket of hot water for 30 minutes, then put the resulting mush into the blender until all the paper has disappeared and you have a grey creamy blob.

3 If you do not have a blender, use a whisk – this needs a bit of elbow grease.

4 If you want to add some colour, put in a drop of food colouring or add a handful of brown or red onion skin (just the papery bit).

5 Put two tablespoons of glue in your tub of water, and add all the paper pulp you just made. Mix it really well with your hands – it's a bit messy!

6 Slide the frame into the bottom of the tub and then lift it very, very slowly (count to 20 while you lift). Let the water drain for about a minute.

8 Using the hottest setting on your iron, steam out the paper.

7 Put the frames out in the sun, or rest them on a tea towel until they are completely dry. Then gently peel off the paper.

And hey presto – you have your very own recycled paper!

UK

Confederation of
Paper Industries

1 Rivenhall Road
Swindon
Wiltshire, SN5 7BD
Tel: 01793 889 600
www.ppic.org.uk

The Directory Recycling Project

*For recycling all those old
copies of Yellow Pages.*
Yell Group plc
Queens Walk, Oxford Road
Reading
Berkshire, RG1 7PT
Tel: 0118 959 2111
www.yellgroup.com
*and click on the Governance
and Directory icon*

UNITED STATES

American Forest and
Paper Association

*Lots of interesting facts about
American forestry and paper
recycling. Accessible and well laid out.*
1111 19th Street, NW, Suite 800
Washington DC 20036
Tel: 202 463 2700
www.afandpa.org

Paper University

*A website for teachers and
pupils to learn about paper,
including recycling facts,
run by TAPPI, the Technical
Association for the Pulp, Paper
and converting Industry.*
TAPPI
15 Technology Parkway South
Norcross, GA 30092
Tel: 770 446 1400
www.tappi.org/paperu

CANADA

Canadian Pulp and
Paper Association

Forest Products
Association of Canada
Suite 410–99 Bank Street
Ottawa, ON K1P 6B9
Tel: 613 563 1441
www.cppa.org

AUSTRALIA

Australian Paper
Industry Council

PO Box 3120
Manuka, ACT 2603
Tel: 61 2 6295 7312
www.apic.asn.au

SOUTH AFRICA

Mondi Paperwaste

*Offer a range of free services
including paper pick-up,
paperbanks and confidential
shredding services.*
PO Box 688
Pinetown, 3600
Tel: 0800 022 112
www.mondi.co.za

Recycle your magazine and seven days later it could come back as your newspaper.

Case study 1

USA

Germany

Japan

The environmental policies of the mega-economies

The United States, Germany and Japan are the world's three richest nations. As such, production, consumption and waste in these countries are high, and their ecological footprint is respectively large. The United States alone produces close to 25 per cent of the world's carbon dioxide emissions, and approximately 720 kilograms of waste per annum per capita! As a precursor to some of the case studies we will be looking at, it is interesting to consider how these countries differ in their approaches to environmental issues.

As it stands at the moment, Germany is a definite leader in all things green. But this hasn't always been the case. In the 1970s the US was much more advanced in its environmental policies. Due to tax breaks afforded to NGOs, there was a powerful movement to regulate the environmental impacts of industry. This led to a backlash during the Reagan era, and the powers of the NGOs were greatly curtailed in favour of 'industrial development'. In the meantime, Germany, who had been far more *laissez faire* during the 70s, with West Germany desperately developing its economy at all costs, suddenly got a reality check when it was pointed out that acid rain was destroying much of its indigenous forestry. The Green Party began playing a more important role in the country's politics, and Germany was among the first nations to commit to domestic greenhouse gas emissions reduction measures.

Japan has always trodden a careful path between the policies of the US and Germany. In the 1970s and 80s, Japan had a strong record in energy efficiency and air pollution but had been criticised for its policies on forest protection and for its fishing and whaling industries. In the early 90s, shortage of landfill space, air pollution caused by traffic congestion, changing weather conditions and a fear of China's possible pollution overspill, led to a reconsideration of their environmental policies, and much more stringent regulations were put in place.

Germany's Green Dot symbol, which can be found in various colours on all types of packaging.

The Situation Today

United States – Despite President Bill Clinton's support for environmental reform, and his agreement to reduce greenhouse emissions by seven per cent in adherence to the Kyoto Protocol, the senate repeatedly blocked his attempts to make significant advances in this direction. The George W Bush administration then took big steps back, by arguing that climate change is an unproven theory and by refusing to adhere to the Kyoto Protocol. The United States is now one of the poorest performers in terms of its contribution to international environmental efforts.

Germany – The combination of a social market philosophy and the strength of environmental groups (including the Green Party) have meant that Germany's eco-policies are among the best in the world. They have achieved the most substantial greenhouse gas emission reductions of any country, have implemented a rigorous set of eco-taxes to regulate industry, and are renowned for their commitment to waste reduction. Their economy has been significantly weakened over the past few years, and it remains to be seen whether this will have a detrimental impact on their environmental successes.

Japan – When the US dropped out of the Kyoto Protocol, Japan made a decision to follow Germany and the EU, and their policies are strengthening. The incentive for change in Japan comes from the government, rather than from the people, as is the case in Germany. This means that their efforts perhaps lack some of the convictions of their European counterpart, but there is no doubt that Japan's policies are rapidly improving. The minister of the environment in Japan writes:

> In the past year, Japan first endured a record-breaking summer heatwave, and then found itself suffering great loss of life and significant damage from an extremely high number of typhoons hitting the mainland. In such circumstances, I believe every citizen directly perceives the changes and abnormalities of the climate system and their consciousness has been raised as a result. This heightened awareness gives us an opportunity to renew our awareness of climate change and other environmental problems as issues deeply connected to us.[1]

America, Germany and Japan are the world's wealthiest countries and therefore have a responsibility for global environment stewardship. Where they lead others will surely follow, and when they set a negative example, an excuse is provided for the rest of the world to follow suit. Their policies will set the tone and as they evolve, so will the future of recycling.

1 Koike, Yuriko, "The Spirit of Mottai Nai", *Our Planet*, vol. 16, no. 1, UNEP Publications

Case study 2

Senegal

Construction workshop in Dakar

Involving students, local women and children to experiment with recycling and self-build techniques

In developing countries, the approach to waste is very different from that of wealthy Western societies. It is harder to take planned obsolescence for granted, when resources are scarce and poverty is only one step away at any given time. In Cuba, waste is cunningly converted into taxi signs and oil lamps (see Case study 7, p. 110). In other places, waste material is actually used in an architectural context; as a basic building material.

In December of 2004 a group of 15 architecture students from the University of Sheffield travelled out to Senegal, in West Africa, to meet a group of women planning to build their own houses. This group is part of the REFDAF (Réseau de Femmes Pour un Développement Durable en Afrique), a network of women's associations, working together for sustainable development in Africa. By using principles of self-organisation, collective design and self-build, the 325 women hope to construct a new community called Cité des Femmes, located in Keur Massar, a suburb of Dakar. When the students began working with REFDAF, the project was already underway, with enough money raised to purchase the first 100 plots of land.

The students and REFDAF worked together on a series of workshops that concentrated on the design of the construction workshop, which REFDAF planned to be the first building on the site. The outcome of a live project earlier in the year has been a database of sustainable self-build techniques to be used in the construction of the future houses.[2]

Students and participants in the project create a tiled roof out of used soft drink cans.

2 "Live project" refers to an institution in Sheffield for shelter in Africa (November 2004). www.shef.ac.uk/architecture/main/galery/gal/diploma/liveproj/lp6/index.html

Below and below opposite: Students help secure bamboo structures that will become an exterior wall to one of the houses. Above opposite: Casually disposed waste became a source material for the students and the women.

The students and the group of women used the database in their first workshop to begin a discussion about the possibilities of using local and recycled materials, craft skills and traditional building techniques. This conversation was continued in a series of design workshops for the proposed construction workshop, where drawings and models were used by the students and women to share design ideas. Finally, a construction workshop took place, involving the REFDAF women and their families and sustainable self-build techniques were used to construct a sand bag wall, with glass jar lights, a tyre wall, and a small pavilion using PVC pipes as columns and flattened cans as roof tiles.

Opposite: Tyres are filled with sand for the foundations, and covered with mud for a boundary wall. Below: The women from REFDAF participating in the project.

The women chose techniques that were more appropriate to them; they were looking for easy to find and cheap materials (sand, rice bags, bamboo, cans, bottles, tyres), valorising their skills (i.e. cutting, weaving and knitting), their aesthetic taste (like the roof covered with can "tiles", which reinterprets the idea of patterning in African textiles) and their informal networks (women from Casamance provided the bamboo, others the rice bags and the cans, etc.). They were also interested in materials and techniques that were easy to build with, allowing women and children to participate together in the construction. The 50 women and children who have participated in the workshop will build the first collective building on the site, training others to get involved in the future construction of houses.

Doina Petrescu, 2005

We of an older generation can get along with what we have, though with growing hardship; but in your full manhood and womanhood you will want what nature once so bountifully supplied and man so thoughtlessly destroyed; and because of that want you will reproach us, not for what we have used, but for what we have wasted... so any nation which in its youth lives only for the day, reaps without sowing, and consumes without husbanding, must expect the penalty of the prodigal whose labor could with difficulty find him the bare means of life.

Theodore Roosevelt, "Arbor Day – A Message to the School-Children of the United States", 15 April 1907

Glass

What is it?

Glass furnaces in the past were much smaller than the vast industrial facilities the industry uses today.

Glass is a common household material made from a mixture of sand, soda ash, limestone and additives. It is created by heating the components to a temperature of around 1600 degrees Celsius in large furnaces. The resulting molten glass is then cooled very quickly, so that the material does not have a chance to crystallise, but remains smooth, solid and brittle.

On a relative scale, glass is not a particular danger to the environment. The raw materials it requires are cheap, plentiful and easy to source, although some do need quarrying. If exposed to the elements, glass will break down into sand, and in landfills, it will not decompose, but neither will it leach toxic substances.

The real drawback is the significant quantities of energy required to heat the furnaces to create the molten glass, and once it is made, the fuel required to transport it. Although the weight of domestic glass has been reduced by 40 per cent over the past 20 years, it is still heavy and transportation results in high carbon dioxide emissions.

Why recycle?

Glass is an ideal product for recycling. It can be melted and reformed infinitely without any loss of quality. One 340 gram wine bottle can be melted and re-formed into a 340 gram bottle with no waste or by-products generated. A glass bottle sent to recycling can be back on the shelves within 20 days.

- Although glass only comprises two per cent of the volume of the average UK household bin, it accounts for seven to eight per cent of the weight of total solid wastes.

- Glass recycling saves about 50 per cent of the energy required to produce virgin glass. These savings come from the slightly lower temperatures required to melt it down (an energy saving of ten per cent) and cutting out the energy required to source and transport the raw materials.

- Due to the lower temperatures required, glass recycling prolongs the life of melting furnaces by up to 20 years.

- Each ton of recycled glass reduces mining waste by 230 kilograms.

One 340 gram wine bottle can be melted and re-formed into a 340 gram bottle with no waste or by-products generated.

Brown glass -
No plastic bottles!
Please use between
8am and 8 pm ONLY
Do not leave
rubbish around
the bins

Green glass -
No plastic bottles!
Please use between
8am and 8 pm ONLY
Do not leave
rubbish around
the bins

Clear glass -
No plastic bottles!
Please use between
8am and 8 pm ONLY
Do not leave
rubbish around
the bins

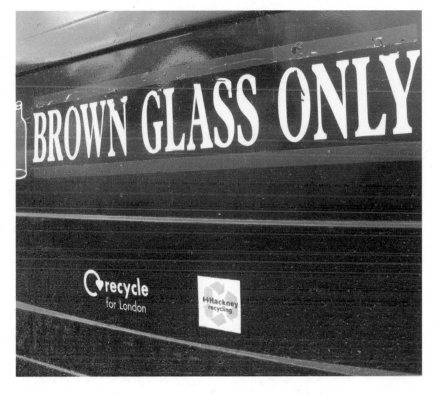

A molten lump of glass is called a 'gob'.

In the past, a glass blower attached a tube to a gob to blow the glass into shape. The larger the gob, the harder they had to blow, and the bigger their cheeks had to inflate.

Today, someone with a big mouth is told they have a big gob.

Problems with glass recycling

There are different types of glass in the waste stream. Container glass (bottles and jars) accounts for 64 per cent of UK glass production. Other types of glass are flat glass (glazed windows), fibreglass, domestic glass (for oven and kitchenware) and special glass (cathode ray tubes, light bulbs, medical and scientific glass). Container glass is the only glass that is being recycled at present. All other types are considered contaminants in the glass recycling process, and as yet, there are very few facilities set up to process them individually.

A barrier to recycling container glass in the UK is the shortage of clear glass. Britain exports clear glass in the form of spirit bottles, and imports green glass in the form of wine bottles. This leads to a surplus of green cullet (crushed glass). Recyclers are finding new ways of reprocessing this green cullet, and 85 per cent of green glass manufactured in the UK is recycled product.

In the US, green cullet is also a problem. In California most green bottles come from the French and European wine industry. Therefore, although the facilities have been set up to recycle the glass, the bottles then need to be shipped hundreds of miles away to be filled – creating a cost and energy inefficient situation. This phenomenon is known as dislocation.

The most serious problem with glass recycling is the contamination that occurs when colours are mixed. Most glass comes in amber, clear or green. Each colour must be recycled separately, or the quality and colour integrity of the final product will not be maintained. Due to its fragile nature, if people do not separate bottles properly, the glass tends to break, and it is very difficult to sort the different coloured shards so they may be recycled. At the moment, the bulk of 'mixed cullet' (crushed, unsorted glass), is relegated to landfills. A new process is currently being developed that will make it possible to decolourise glass, so that it can be recycled as clear glass, or recoloured green or amber. However, this technology is not yet in use, and mixed cullet remains a problem.

Top: A selection of container glass.
Middle: Special glass, such as that in light bulbs is not suitable for recycling.
Bottom: Clear cullet before and after it is ground into powder.

The most serious problem with glass recycling is the contamination that occurs when colours are mixed.

Another challenge facing the glass recycling industry is the increasing dominance of plastics in the packaging market. Less than a generation ago, most food packaging was made of glass. Now it is being replaced by cheaper and lighter PET plastics (see Plastics section). Between the years 1994 and 2000, glass bottle shipments to the United States plummeted from 3.6 billion units to 800 million units, whilst PET bottle shipments doubled to 25.6 billion units. With less product to recycle, the financial potential of glass recycling decreases.

The recycling figures in Britain and the United States are pretty poor, with each country recycling approximately 30 per cent of container glass. In Switzerland and Finland, more than 90 per cent of glass is recycled.

At one time, it was commonplace to return bottles to the retailer for reuse. This is obviously the most efficient form of recycling as no energy is required at all to melt the bottles down.

However, the transport of bottles for refilling stopped becoming financially viable as beverage manufacturers began to opt for bigger plants further apart.

In Britain, milk bottles are still being reused. You may notice that the bottles are heavier than most other glass containers. This is so that the bottle will withstand cleaning and reuse an average of 20 times without breaking.

In Austria there is a facility where you can take a glass milk bottle from home and refill it yourself at a pump, which tells you which farmer it came from and how many hours old it is.

The recycling process of glass

3
Large contaminates are removed (chunks of ceramic or plastic).

2
They are transferred into one of three big hoppers – one for clear glass, one for amber and one for green, and passed from there onto a conveyor belt system.

1
Glass bottles are delivered and stored in loading bays according to their colour until they are ready to be processed.

10
It is then moulded or mechanically blown into new bottles or jars.

9
The glass is mixed with raw material in a furnace and melted down.

4
The glass is then crushed and screened to remove more large contaminates – such as wine corks and paper. The crushed glass is known as cullet.

5
From there it is passed through a big magnet which detects and removes bits of metal – bottle tops and wire, and another screen which detects little bits of stone and ceramics.

6
The cullet is then vacuumed up and screened yet again for paper and non-ferrous metals (aluminium foil).

7
In the final step of the screening process, the glass is run over by a laser beam. It passes through the glass, but any remaining contaminates will deflect the beam. These contaminates are sought out and removed.

8
The glass cullet is then transferred to the glass factory.

How do I recycle glass?

The most important rule of thumb is to clean and colour-sort your glass. You don't need to scrub bottles and jars, a rinse is sufficient, and paper labels do not need to be removed. Big foreign materials (such as bottle tops, corks and stoppers) should be removed, as they can get caught in the production line and cause problems.

Not all glass is recyclable. Most areas only have facilities for recycling container glass. This is changing, and schemes are slowly being set up to recycle other types of glass waste. However, for the time being, window panes, light bulbs, laboratory glass, ovenware and mirrors, each of which have different chemical compositions, are generally not suitable for recycling.

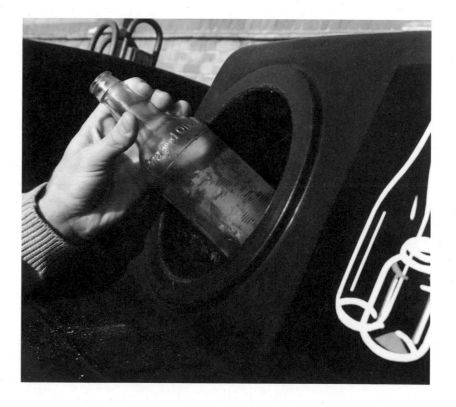

A few tips

- Blue bottles are recycled as green. If you use a bottle bank, put your blue glass in the green bank.

- If a bottle is returnable, return it, rather than recycling. Reuse always saves more energy than recycling.

- Recycle all glass containers, not just wine and beer bottles. Ketchup bottles, jam jars, mayonnaise jars and juice containers, to name but a few, are all recyclable.

- If a bottle has been colour coated, this is still recyclable, as the coating can be burned off in the furnace. Check the top and bottom of the bottle to find out the original colour and put the bottle in the appropriate bank.

- Store your bottles indoors until you are ready to recycle them. This reduces moisture contamination and mould.

Glass recycling is one of the most efficient recycling processes and is therefore promoted around the world.
Below: London's Recycle Now campaign.
Below right: A bottle bank in Frankfurt.

What happens after it is recycled?

Container glass that has been properly sorted and decontaminated can be recycled as container glass. However, contaminated and mixed cullet are not suitable for reuse as container glass. This glass can be recycled into other forms:

- 'Glass grit' can be used for grit blasting.

- Powdered glass can be used as a 'fluxing agent' for brick and tile manufacture.

- Glass can be ground down into processed sand, which can be used as sports turf or as a sand replacement in concrete and cement production.

- It can be used as an aggregate in road paving – a material known as Glasphalt.

- It can be incorporated into the fibreglass manufacturing process.

- Water filtration uses sand as a filter. At the moment it is not ecologically beneficial to use glass as a replacement water filter, but technologies are being developed to make this application more efficient.

Mixed cullet is unsuitable for recycling into container glass.

Although these alternatives are preferable to landfilling glass, they are not genuine recycling, and many of the environmental benefits of glass recycling are reduced if old bottles are simply replacing sand.

UK

British Glass
9 Churchill Way
Chapeltown
Sheffield, S35 2PY
Tel: 0114 290 1850
www.britglass.org.uk

*The children's section of
the British Glass website
can be found on:*
www.recyclingglass.co.uk

Glass Forever
*A glass recycling
website geared towards
kids and teachers.*
Jo Hollins
Rockware Glass Ltd.
Headlands Lane
Knottingley
West Yorkshire, WF11 0HP
Tel: 01977 674 111
www.glassforever.co.uk

UNITED STATES

Glass Packaging Institute
*Everything you need to know
about glass, glass packaging
and glass recycling, with
links to local resources.*
740 East 52nd Street
Indianapolis, IN 46205
Tel: 317 283 1603
www.gpi.org

Container Recycling Institute (CRN)
*For plastic and glass bottles and
cans recycling facts and figures.*
1601 N. Kent Street, Suite 803
Arlington, VA 22209
Tel: 703 276 9800
www.container-recycling.org

CANADA

Alberta Bottle Depot Associaton
*Specifically for glass
collection in Alberta, but
good general information.*
202, 17850 105 Avenue
Edmonton
Alberta, T5S 2H5
Tel: 780 454 0400
www.geocities.com/
RainForest/Vines/6156

AUSTRALIA

Australian Glass and Glazing Assocation (AGGA)
6 Kinkora Road
Hawthorn
Victoria 3122
Tel: 03 9853 3464
www.glassandglazing.com.au

SOUTH AFRICA

Glass Recycling Association
PO Box 562
Germiston, 1400
Tel: 011 827 4311
www.glassrecycling.co.za

Case study 3

Brazil

Curitiba: The importance of town planning

Curitiba, a small city in southern Brazil with a population of 1.6 million, is held as a worldwide paragon of urban planning excellence. With limited funds at its disposal it has created an uber-efficient transport system based around buses, a clean and pedestrianised city centre and an accessible, first-rate education system. It has addressed issues of poverty by rehabilitating poorer areas of the city, and to top it all it boasts 52 square metres of green space per inhabitant.

It also leads the world in its waste management policies, recycling close to 70 per cent of its rubbish. The process starts with acknowledgement of the realities of the city. The favelas, the slum areas, are chaotic places with steep, narrow, unpaved roads. Rather than sending in lorries to try and collect rubbish and hunt down clandestine waste dumps, the inhabitants are 'paid' for their rubbish. In other words, they bring the contents of their bins to community centres where it is exchanged for bus tickets or local produce (thus helping local farmers). 60 kilograms of rubbish earns 60 tickets, which can be exchanged for enough food to feed an entire family for a month. This policy encourages 'informal collection' – people actually want to collect waste; when it has a real value, rubbish becomes a desirable commodity. Similar exchanges of trash for goods occur at schools and factories, with food, notebooks and toys provided for diligent recyclers.

The sorting of rubbish also addresses a number of problems by providing jobs for the poorer population. Curitiba's citizens separate their rubbish into just two categories – organic and inorganic. This is picked up by trucks according to its type, and taken to a plant (itself built of reclaimed materials), where it is sorted by recent immigrants, recovering alcoholics and disabled people. The recovered materials are then sold on to local industries; styrofoam is shredded to stuff quilts and compost is used for regional agriculture. These

The entire waste management programme works out the same as landfill costs.

efforts are supplemented by a programme called 'All Clean', in which unemployed or retired people are hired on a temporary basis to clear litter from hard-to-reach land in the parks and by the rivers. The entire waste management programme works out the same as landfill costs.

How has this system evolved and why does it work so well?

Curitibans have a lot to thank Jaime Lerner for. As a young architect in the 1960s, he devised a masterplan drawn around main arteries of transport for the city. Realising that architects have little clout in the political sphere, he became mayor – a position he has held on and off for the past 30-odd years. Then he simply implemented his plan. The clear flow of transport gave a structure to the development of the town and one success led to another, leading to a virtuous circle based on mutual respect – the city provides its residents with a pleasant environment, the residents respond by looking after and taking pride in their city. In an article for *Our Planet*, Lerner writes:

> The potential of local-level action is self evident. Consider, for instance, the link that exists between local action and external debt. Brazil's debt inhibits the country's development: this is true for many other countries as well. However, if each city were to act against wastefulness, some of the energy-generating resources currently purchased from abroad would be no longer needed.[3]

Obviously this is an easier philosophy to implement in a town of less than two million inhabitants, than it is for a megalopolis like London or New York, but there is a fundamental truth at the heart of this statement: waste is a resource that everyone has access to. And it's up to the government and to each and every citizen to realise its value. Lerner writes:

> Authorities must not fall into the 'tragedy syndrome', where problems seem insurmountable and where the inhabitants become complacent, believing their isolated actions to be insignificant. If responsible action is encouraged now, the community will respond positively to our appeals in future. Above all, if the inhabitants feel respected, they will respect the environmental issues presented to them.[4]

3 Lerner, Jaime, "Change Comes From the Cities", *Our Planet*, vol 8.1, June 1996
4 Lerner, "Change Comes From the Cities"

Case study 4

The Recycling Lottery

Norway

Squash it, write your name on it and win!!!

This is the slogan of Norway's Recycling Lottery, a nationwide scheme designed to encourage consumers to recycle used drinks cartons. It is a slogan familiar to Norwegian people; the lottery is advertised on television and billboards, promoted in newspapers and through leafleting. It is also literally what Norwegians must do if they wish to enter the lottery. Participants place six empty, squashed beverage cartons inside a seventh to form a 'cube'; they then write their name and a contact telephone number on the 'cube' and either take it to a recycling bin or wait for it to be collected by the local authority. From each municipality a certain number of cubes, proportional to the population, are put forward to the national draw and the recyclable products are effectively the ticket to win a cash prize.

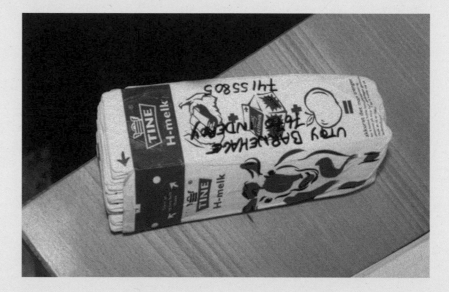

Each year Norwegian consumers produce roughly 500,000 tons of packaging waste. This waste contains approximately 19,050 tons of used beverage cartons and approximately 30,800 tons of other packaging. Whilst this waste represents a relatively small fraction of the total household waste produced in Norway (about two and a half per cent of the total of 1,900,000 tons) the Norwegian government has taken serious steps to ensure that post-consumer waste is being recycled rather than condemned to landfill.

During the 1990s the government set big industries and manufacturers recycling and recovery targets. These targets are still in place, and manufacturers incur heavy fiscal taxes on their products if their performance is not up to scratch. Similarly to what happened in Germany with the Green Dot programme (see Case study 6, p. 106), each waste stream in Norway set up its own private, not-for-profit Producer Responsibility Organisation (PRO). These organisations are responsible for ensuring that manufacturers

By 1996, despite a convenient collection scheme, less than 35 per cent of the population were recycling cartons.

take responsibility for their own waste production. The fiscal taxes far outweigh the cost of recycling so that it is in the interests of producers to ensure that government targets are met. Returkartong (literally Carton Recycling) is the 'PRO' responsible for ensuring that packaging recycling targets are met. They found a viable market for this post-consumer packaging, the empty cartons are sold to a paper mill that recycles them into high quality paper products. They also arranged for municipal authorities to make house to house collections of waste beverage and packaging cartons.

By 1996, despite this convenient collection scheme, less than 35 per cent of the population were recycling cartons. The Recycling Lottery was set up in 1997 as a way of solving this problem and offering an incentive to those who recycle. In the first year an annual draw was made with prizes totalling 2 million kroner (£162,000 or $300,000). Any person or organisation that squashed, cubed and signed their recyclable cartons was eligible to enter and stood the chance of winning a top prize of £81,000 ($145,000). The scheme has now been altered so that there are quarterly draws with a top prize of £14,000 ($25,000) and smaller prizes of £1,400 ($2,500), but the total of 2 million kroner and the general principle is still the same.

Utoey Nursery School was the first lottery winner in 1998, and used the money to buy school equipment. Here the children celebrate with their teachers.

The scheme was heavily promoted and has been largely successful. After the introduction of the lottery, around 50 to 60 per cent of the population were recycling cartons. That figure has now increased to 68 per cent. The prize money is relatively small beans compared to the 4 million kroner prizes given out each week by the Norwegian National Lottery and Returkartong say that it is not a central part of their strategy. However, the Recycling Lottery is a way of reaching out to consumers who are less environmentally aware, offering an incentive to those who recycle. Continued coverage on television and in local and national newspapers raises awareness of the lottery and, by extension, recycling. Both the 'PRO' scheme and the lottery appear to have been successful in Norway. Municipal authorities now collect recyclable materials from roughly 80 per cent of the population and around 90 per cent of Norwegian citizens are currently aware of the country's recycling system.

The Frisk Boys 91, under 14s hockey team proudly holding up their cheque.

Case study 5

Zurich: Thinking twice about the garbage

Switzerland

The most notable thing about garbage in the picture-perfect medieval city of Zurich, is that it's nearly impossible to find any.

On Untere Zaune, a winding cobbled street in Zurich's historic old town, household trash is collected just once a week – with the precision of a military strike, from seven to nine a.m. on Fridays. Just after seven, the kerbs are busy with residents and business employees putting out a white plastic Zuri-Sack or two.

This is the city's official, and costly, trash bag, which must be purchased from the government at about five Swiss francs ($4.25 or £2.50) apiece, depending on size. It seems impossible that, in a whole week, humans living in the twenty-first century have produced so little refuse.

"I have a special room where I keep garbage until it's the right day", said Marianne Schlaepfer, a Christian minister at a hospital, leaving a Zuri-Sack by the kerb across from an ancient church. "The programme is tough but you get used to it. It does make you think more about garbage and the environment."

In Europe, household waste is increasing by ten per cent every five years despite calls by nearly all governments to reduce it. At current rates, the amount of paper, glass and plastic waste will be up by as much as 60 per cent in 2010 compared with the 1990s, according to the Association of Cities and Regions for Recycling, or ACRR, a non-governmental organisation in Brussels. But here in Zurich, as in a few other parts of Europe, draconian disposal and recycling programmes in place for more than a decade have dramatically reversed the trend.

Household trash in Zurich has decreased overall by 40 per cent since 1992, said Alfred Borchard, logistics manager of the city's Waste and Recycling Bureau. Zurich power plants that used to burn local trash to make electricity now burn garbage from Germany instead.

"They've turned us all into recyclists!" said Florian Eidenbenz, 40, a sound engineer who on a recent day rode his bicycle a kilometre to a recycling centre by the lazy Limmat River in the centre of town to dispose of his family's bottles.

Separating them by colour, since fines are high, Eidenbenz added: "You learn to cope because it reduces waste and garbage. I mean it's clever: when they charge so much for Zuri-Sacks, you think twice putting things into the garbage."

Zurich started charging for garbage bags in 1992. Within a year, household trash began decreasing from 140,000 tons a year citywide to 100,000, today's level.

In a survey by the ACRR of several dozen cities in northern Europe, only a few small cities in Austria and Holland produced less trash per person. The average Zurich family produces just one Zuri-Sack a week.

To achieve this, residents like Schlaepfer have learned to live with complicated incentives for people not to create garbage, from the mandatory use of Zuri-Sacks for trash disposal to a byzantine collection schedule that allows paper and cardboard to be discarded only once a month each (on different days), and most other types of trash even less often.

In Paris, Rome and New York, where the rumble of garbage trucks is like background music and collection happens frequently, it is easy to toss things casually into the trash can. Zurich, in contrast, has made throwing things away nearly impossible — and also costly.

It is hard to compare figures on household waste, since countries tend to "measure it differently", said Francis Radermaker, executive director of the Brussels based Recycling Association. The average European generates 540 kilograms, or 1,190 pounds, of waste a year, according to the Wuppertal Institute, a German environmental research organisation, compared with just over 400 kilograms per capita in Zurich.

Trash generation tends to be tied in a loose way to a country's economic development, with Americans producing the most trash in the world, and poor African nations the least. "Trash increases with the standard of living," Radermaker said, "and so even though people are more conscious than they used to be, they are not conscious enough."

But statistics understate the full impact of Zurich's programme, which has revolutionised the way consumers shop and the way megastores do business. Since Zurich residents had no simple way to dispose of packing materials under the new programme, for example, they became reluctant to buy products like televisions and microwaves that normally come swathed in layers of cardboard, plastic and Styrofoam.

Stores like IKEA now put products in far less packaging, or allow customers to unwrap at the store, and appliances now leave stores nearly naked. "Packaging is minimal," Borchard said. "People didn't want to take anything extra home."

With less garbage being produced, the city reduced collection schedules from daily to twice weekly and then to once a week for most neighbourhoods. When you have to live with your trash for days or weeks (depending on type), garbage naturally becomes a bit of an obsession.

Like many people here, Schlaepfer, the hospital minister, has set up a miniature trash-and-recycling facility – in her beamed fourth-floor walk-up apartment, which looks out on the river and church steeples of Zwingli Platz. She has devoted a small room next to the kitchen to trash. There is one rack for storing newspapers and another set of shelves for keeping cardboard boxes, since these are picked up only once a month. There is a bin to hold household trash for the Friday pickup, and another that stores bottles and cans and textile garbage, which Schlaepfer says she takes to a collection centre every few weeks.

The key to the operation sits in her kitchen drawer with the telephone books: Zurich's master refuse collection schedule, listing which type of garbage will be picked up in which neighbourhood on which day.

David McQuillen, an American who lives in Zurich, said that he, for one, can never keep track. "I of course lose this calendar the day after I get it. And then I just wait until I see cardboard being put out by my neighbours – the Swiss are so good at this – before doing the same."

In this famously organised and law-abiding country, the Swiss seem to have accepted and absorbed the complexities as second nature. After all, this is a city where neat little yellow dispensers adorned with the silhouette of a schnauzer provide dog walkers with free pooper-scooper bags.

It is perhaps not surprising that another country known for exemplary trash programmes is Japan, where culture and lack of space have served to reinforce regulations, said Guido Sonnemann of the United Nations Environment Programme. "In Japan, there is a long tradition that there are things you do because they are good for the community, even if it's inconvenient for you personally."

Though recycling laws now exist in virtually all European countries and bins now dot sidewalks in Paris, London and Rome, compliance is variable, especially in southern Europe.

Zurich's programme has spread throughout Switzerland over the last two years. Many other European countries are implementing programmes of their own, with varying strategies. A number in northern Europe have reduced the frequency of trash collection dramatically. In Finland, the garbage man comes only once every two weeks in many cities. But that may not be practical in a place like Italy, experts say, where the climate is hotter, apartments are smaller, and where food waste might rot in that period.

Perhaps the biggest innovation is a 'pay-as-you-throw' policy, where garbage collection fees are linked in some way to the actual amount of trash a household produces. In much of the world, garbage taxes are a flat fee or related to household size.

In Zurich, an innovation as simple as the Zuri-Sack serves as a kind of garbage tax, discouraging trash. In Germany and Holland, many cities have adopted more high-technology 'pay-as-you-throw' plans.

In Dresden, microchips in dumpsters measure the volume of garbage, which is then linked to charges on households. Some cities in Denmark already use such readings as a sort of 'garbage meter', following the same principle that gas and electric companies use to calculate bills.

In studies, 'pay-as-you-throw' fees are "the most likely to change household behavior patterns", according to the ACRR. In the past couple of years, they have become widespread in Austria, Belgium, Finland, Luxembourg, Sweden and Switzerland.

Plenty of people in Zurich acknowledge that it was less public-spiritedness than the city's strict enforcement system and the threat of big fines that forced them into eco-consciousness.

The Zurich trash programme employs teams of inspectors who sift through sacks of illegal household garbage, looking for clues as to the culprit. Fines are as much as 260 Swiss francs (£115 or $206) – and that for a first offence. "Many of us, including me, ran into a fine at some point," Eidenbenz, the sound engineer, said. "And that was highly educational – especially for stingy people like me."

Elisabeth Rosenthal
Courtesy of: International Herald Tribune,
23 April, 2005

备战、备荒、为人民。

大力回收废钢铁
及其它废旧物资！

每回收**10**吨废钢

可炼**9**吨好钢

— 可制造

北京130载重汽车4辆

手扶拖拉机12台

自行车300辆

可节约 —

生铁10吨

每炼1吨生铁需要:

铁矿石2吨

石灰石0.5吨

煤1吨

内部张贴

北京市物资回收公司

Sustainable development is development that meets the needs of the present without compromising the ability of future generations to meet their own needs.

Commission on the Environment and Development

Metal

Scrap metal is divided into two types; ferrous (iron and steel) and non-ferrous (everything else). Ferrous metals are the most recycled materials around the world, with over 69 million tons of steel recycled last year in the United States alone. Non-ferrous metals are all those that are not derived of iron or steel and include aluminium, copper, gold, silver, brass lead, magnesium and mercury. When we talk about metal recycling, we are usually referring to steel and aluminium, as the other metals are either found in minute quantities, or, in the instances of gold, silver and brass, have a high enough value that they are not readily disposed of.

The main problem with all metals is that they are mined, and all types of mining have a devastating effect on the environment. Pit mining, required to source steel, leaves disfiguring scars across the landscape, and surface mining, used to extract aluminium, employs vast quantities of poisonous chemicals, which tend to leak into the ground, contaminating the earth and water for miles around.

Scientists point out that it takes many decades for water-soluble chemicals to percolate through the soil and pollute water sources. Therefore, we may not even realise the extent of the damage we are doing, as the toxins in water sources today may be the result of industrial activities from a generation ago.

The good news is that unlike paper or textiles, metals are 100 per cent recyclable, i.e. they can be recycled without losing any of their properties. This is a big incentive for industries to treat used steel and aluminium as a primary source, rather than as a waste material that needs to be managed.

Below: The devastating scars of strip mining. Bottom: Greenpeace activists protesting such mining methods are dragged off by the authorities.

Steel

Steel is one of the few materials you can purchase that is guaranteed to contain recycled product. In every steel can or new car you buy, a minimum of 25 per cent of the product will be made of recycled steel. In the Western world, recycled steel makes up on average 40 per cent of the content of new steel, and 70 per cent of steel scrap is recovered. Virtually 100 per cent of cars on the road today will eventually be recycled (although only 75 per cent of the material within them is recyclable).

There are a number of reasons for these encouraging statistics. First and foremost, it is very cost effective to recycle steel. Due to its magnetic qualities, it is easy to sort, melt down and reuse. Secondly, due to its relative efficiency, there is a long history of steel recycling. In North America, the steel industry has been recycling scrap for over 150 years. This means that the technology for processing scrap steel is extremely advanced. In fact it takes a grand total of 45 seconds to shred a car into fist sized pieces of steel for recycling!

Cans and cars are two of the most common domestic uses of steel.

Virtually 100 per cent of cars on the road today will eventually be recycled.

In the developing world, the recycling statistics are much lower, as the stock of steel in the economy is essentially fixed.

> In countries in the early stages of industrialisation, the creation of infrastructure (factories, bridges, high rises and transportation) leaves little steel for recycling.[5]

5   Brown, Lester, *Eco Economy,* Norton & Company, 2001, p. 136

Every ton of steel recycled saves:

1,150 kilograms of iron ore

650 kilograms of coal

55 kilograms of limestone

1.28 tons of solid waste

As well as reducing:

Air pollution by 86%

Water pollution by 76%

Water consumption by 40%

Energy consumption by 75%

How is steel recycled?

2
In the case of the BOF, virgin molten iron is added to the scrap.

1
Scrap steel is fed into a large furnace or converter. There are two types of converters; the basic oxygen furnace (BOF) requires a minimum of 25 per cent recycled steel and is used to make products such as cans, car bodies, steel framing and steel drums. The electric arc furnace (EAF) is used to make items such as structural beams, bridge spans and steel plates and uses 100 per cent recycled steel.

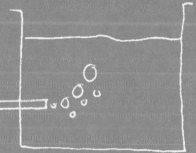

3
A water cooled lance
blows high-purity
oxygen into the metal
to remove impurities.

4
The converter is heated
to about 2000 degrees
Celsius. In about half
an hour, the iron and
used steel are refined
enough to make new,
high quality steel.

5
The hot steel is poured
out of the converter,
cast into solid slabs
and rolled into coils.

What can I do?

Depending on the steel product, there are different ways to approach the recycling process:

Recycle your car: The junkyard is where the recycling process begins. Even the most beat up car can still contain some valuable working parts.

Steel cans: You may not realise the quantity of steel cans you use per day. Pet food, sauces, soups and shoe polish, to name but a few, are all packaged in steel. If you have any doubts about whether a product is steel or aluminium, put a magnet on it. If it sticks, it's steel.

- Contact your local authority to ensure that they have the correct facilities for recycling steel. If they do not run a kerbside scheme, you may have to take your cans to your local depot.

- Make sure your cans are empty.

- In the case of paint cans, try to use up all the paint. If there is paint left over that you do not want to store, contact your municipality to see whether they provide a household hazardous waste service.

- Aerosol cans can be recycled, but they too should be thoroughly emptied before recycling.

Dispose of household goods containing steel (such as refrigerators, washing machines, etc.) in a responsible manner (see section on Household Goods).

Aluminium

Aluminium is the third most abundant element in the earth's crust after oxygen and silicon, and is extracted from an ore called bauxite. Bauxite is found near the surface, in a wide belt around the equator, and is extracted using open-cast mining. This process produces thousands of tons of waste rock, and causes damaging soil erosion. This is particularly worrying as a significant amount of bauxite mining occurs in the Brazilian rainforest, destroying swathes of ancient vegetation.

It is not just the mining itself which impacts on the environment, but the chemicals used to extract the ore. Lester Brown writes that:

> For each ton of aluminium produced, a ton of 'red mud' – a caustic brew of chemicals – is left after the bauxite is extracted. This red muck is left untreated in large biologically lifeless ponds, eventually polluting both surface and underground water supplies.[6]

The final and possibly most worrying aspect of aluminium production is the inordinate quantities of electricity required to smelt it. Aluminium is only found as a compound called alumina – a material consisting of aluminium combined with oxygen. In order to be used, the oxygen has to be separated from the metal, using a powerful electric current.

> Worldwide, the aluminium industry uses as much electric power as the entire continent of Africa.[7]

Recycling one (1!) aluminium can saves enough energy to run a television for three hours or a 100 Watt bulb for 20 hours.

The use of aluminium is on the rise. It is a lightweight material, a good insulator and rust-proof, and is therefore beginning to find its way into traditionally steel-built products such as aeroplanes and cars, as well as comprising the packaging for a whopping 75 per cent of all canned drinks. It is therefore essential that we take aluminium recycling seriously.

6   Brown, *Eco Economy*, p. 129
7   Brown, *Eco Economy*, p. 129

Opposite: Molten aluminium being poured.
Above: The aluminium is rolled into sheets.

Recycling one (1!) aluminium can saves enough energy to run a television for three hours or a 100 Watt bulb for 20 hours.

Alupro is a British non-profit organisation working to ensure that aluminium recycling targets are met.

Some recycling facts

There are two types of scrap aluminium: new scrap — which arises from the manufacture of aluminium products; and old scrap. New scrap has a recycling rate of virtually 100 per cent, but domestic recycling rates for aluminium remain low. In Britain recycling rates are around 40 per cent for cans and 11 per cent for foil.

- Four tons of bauxite are required to produce one ton of aluminium.

- Aluminium is the most valuable recyclable material. It is worth ten times as much as steel. In fact, if all the drinks cans in the UK were recycled through 'Cash For Cans', the revenue would exceed £30 million ($55 million).

- Used beverage cans can be back on the supermarket shelves within six weeks of collection for recycling.

- Sweden is the most efficient beverage can recycler in Europe, recycling 92 per cent of all cans.

Recycling aluminium saves:
95% of energy
95% of carbon dioxide emmissions

The recycling process for aluminium cans

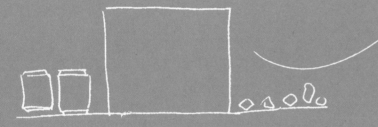

1
The aluminium cans are shredded into small squares and passed through a magnetic drum to remove all steel content.

2
De-coating: The lacquer of the can labels is removed by blowing hot air through the shreds on a conveyor belt. The energy required for this process is also recycled: the hot exhaust from the de-coater is combusted in an after-burner, and this passes over a heat exchanger, which heats the fresh air going into the decoater.

6
The ingots solidify and are rolled into sheets, from which new cans are produced.

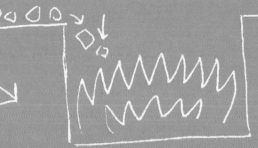

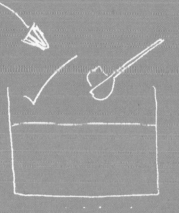

3
The aluminium is then fed
into a massive furnace, in
which a vortex is created,
so that the aluminium
is melted and stirred
at the same time.

4
The molten metal is
transferred to a holding
furnace, where the alloy
composition is checked
and the metal treated to
remove any contaminates
before casting.

5
Ingots are cast by pouring
the molten metal into
a vertical casting unit.
Again, rigorous tests are
performed to check the
purity of the metal.

What can I do?

Aluminium foil and aluminium cans are made of different alloys and must be collected separately. Foil recycling facilities are limited, and there may not be one near you. However, most municipalities run aluminium can recycling schemes, and some will even run 'Cash For Cans' schemes, so you can earn money as you recycle!

Cans: You can crush cans to save space in your recycling box. However aerosol cans, which are also recyclable, should not be crushed or punctured. If you're not sure whether a can is aluminium, test it with a magnet. Aluminium is not magnetic.

Trial schemes have indicated that collecting cans in offices can make 20 per cent savings on the office waste management bill.

In Brazil, aluminium recycling has become a major source of employment.

Brazilians making a living out of collecting and recycling cans earn $200 (£115) per month, as opposed to the minimum wage of $81 (£47).

Encourage an office recycling scheme for aluminium cans. Contact whoever deals with your normal waste and see if they run a recycling scheme. If they don't, they may be able to refer you to someone who will, or contact the local municipality. Certain trial schemes have indicated that collecting cans in offices can make 20 per cent savings on the office waste management bill.

Foil: Aluminium foil is recycled to make cast components for the car industry. You can recycle backing trays, kitchen foil and foil milk bottle tops. Crisp packets are not made of foil. If in doubt, scrunch up the material. If it springs back, it is not foil!

Opposite: Foil containers are often more difficult to recycle than cans.
Right and Below: Local authorities encourage the recycling of this precious resource.
Below right: Crushed aluminium cans, ready for recycling.

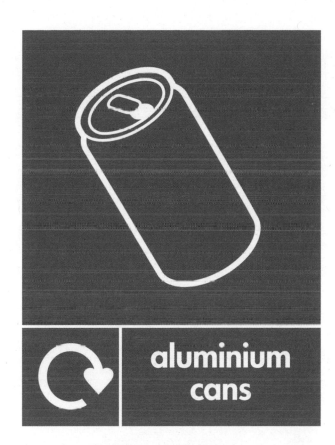

aluminium cans

Right: Crushing cans can save space in your recycling bin.
Far right: An aluminium pot — one of many uses of aluminium.

Womack and Jones trace the origins and pathways of a can of Cola.

"Bauxite is mined in Australia and trucked to a chemical reduction mill, where a half hour process purifies each ton of bauxite into a half ton of aluminium oxide. When enough of that is stockpiled, it is loaded on a giant ore carrier and sent to Sweden or Norway, where hydroelectric dams provide cheap electricity.

After a month long journey, it usually sits in the smelter for as long as two months.

The smelter takes two hours to turn each half ton of aluminium oxide into a quarter ton of metal in ingots 10 metres long. These are cured for two weeks before being shipped to roller mills in Sweden or Germany. There each ingot is heated to nearly 900 degrees Farenheit and rolled down to a thickness of 1/5 centimetre.

These sheets are coiled and transported to a warehouse and then to a cold rolling mill in the same or another country where they are rolled tenfold thinner, ready for fabrication. The aluminium is then sent to England, where sheets are punched and formed into cans, which are then washed, dried, painted with base coat and then painted again with specific product information.

The cans are next lacquered, flanged, sprayed with protective coating to prevent the cola from corroding the can and inspected."[8]

If you think of all the products we take for granted travelling a similar journey around the world, it puts resource depletion in profound relief.

8   Hawken Paul and Amory B Lovins and L. unter, *Natural Capitalism*, Earthscan Publications, 2002, quoting Womack, James and Daniel Jones, *Lean Thinking*, 1996

Other Metals

Other metals are less common, and therefore fewer recycling facilities have been set up to deal with them. However, there are a few things to look out for: copper in electrical wires and plumbing tubes is valuable and recyclable – do not mix the higher grade electrical wiring with the lower grade plumbing tubes. Brass scrap (such as old taps) can be melted down and reused. Lead is found in products such as x-ray materials and batteries (see Batteries section). The recycling processes for lead tend to be product specific, and should be clarified individually.

Top: Taps and coathooks are one of the common uses for brass in the household. Bottom: One of the most common uses for copper is within wiring.

UK

Alupro
For aluminium packaging recycling.
1 Brockhill Court
Brockhill Lane
Redditch, B97 6RB
Tel: 01527 597 757
www.alupro.org.uk

Think Cans
Formerly Alcan – for recycling aluminium cans and promoting 'Cash For Cans' schemes.
Contact via website only
www.thinkcans.com

SCRIB
Steel Can Recycling
Information Bureau
Port Talbot
South Wales, SA13 2NG
Tel: 01639 872 626
www.scrib.org

UNITED STATES

Aluminum Association
900 19th Street NW
Suite 300
Washington DC 20006
Tel: 202 862 5100
www.aluminum.org

Steel Recycling Institute
680 Andersen Drive
Pittsburgh, PA 15220–2700
Tel: 412 922 2772
www.recycle-steel.org

CANADA

Canadian Steel Producers Association
Suite 407
350 Sparks Street
Ottawa, ON K1R 7S8
Tel: 613 238 6049
www.canadiansteel.ca

AUSTRALIA

Cansmart
The steel can recycling council.
PO Box 1854
Wollongong
NSW, 2500
Tel: 1800 073 713
www.cansmart.org

SOUTH AFRICA

Collect a Can
For the promotion and recycling of steel cans. This organisation has been running for ten years.
5/13 Forssman Close
Barbeque Downs, Kyalami
Tel: 011 466 2939
www.collectacan.co.za

NEW ZEALAND

Steel Can Recycling Campaign
Contact via website only.
www.steelcans.co.nz

Case study 6

Green Dot Programme — a lesson in producer responsibility

Germany

Germany is a small, heavily populated country with a strong economy. The combination of these factors means that landfill space is a major problem. In 1991, Germany suffered a garbage crisis of such severity, that it was required to ship its waste over to France. Aware of the political and economical complications this could give rise to, Germany began to take the issue of waste management very seriously indeed.

Since its implementation, the Green Dot programme has led to a 14 per cent decrease in per capita consumption of packaging.

Between the years 1991 and 1993, the country put into place a number of ordinances for the avoidance of packaging waste. According to these laws, distributors and manufacturers — both domestic and foreign — must take full responsibility for their packaging waste, and reuse and/or recycle all materials utilised. In practical terms this means that manufacturers and distributors are required to pay for and facilitate the return and recycling procedure of every aspect of their product's packaging. This could pose serious logistical and cost issues to the manufacturers, and so an alternative was provided. Rather than industries arranging everything themselves, they can pay a fee to the national waste management service; a privately run, not-for-profit system known as the Duales System Deutschland (DSD), and in return for these fees, DSD allows these companies to place its trademark 'Green Dot' on their products.

The fees charged by DSD are proportionate to the packaging used. According to the DSD website, it "depends on the material, weight and volume or area of the packaging. As such it is an incentive to optimise packaging: the less the pack weighs, the lower the license fee will be."[9]

9   www.dsd-ag.de

Packaging is divided into three categories:

Transport – packaging used to ship goods to retailers (crates, pallets etc.).

Primary packaging – the package containing the product (tins, jars, plastic wrappers).

Secondary packaging – any additional packaging to prevent theft or advertise the product.

Transport accounts for one third of the packaging waste stream, and primary packaging about two thirds. Secondary packaging accounts for less than one per cent.

Der Grüne Punkt –
Duales System Deutschland AG

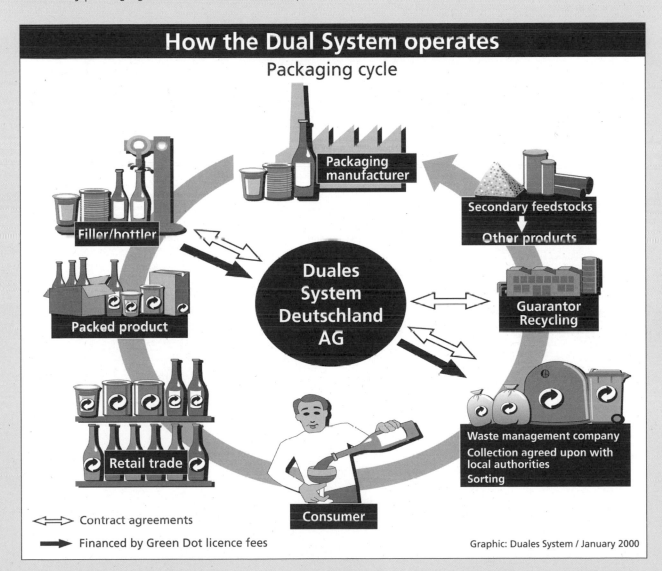

Since its implementation, the Green Dot programme has led to a 14 per cent decrease in per capita consumption of packaging.

On a domestic level:

DSD runs two systems of recycling. Firstly there is a kerbside collection system; residents are provided with blue bags or bins in which they can dispose of containers (juice and milk cartons) and packaging, and yellow bags or bins for metal and plastic. And secondly there is a drop off system (called a 'bring system') – in which consumers take their glass and paper packaging to containers or recycling stations. These are well organised and there is typically a disposal unit located every few blocks in urban areas.

A hefty deposit has also been levied on cans – 25 cents per can. This is returned to the customer when the cans are brought back. Counting the deposit, drinks in refillable containers are now cheaper than drinks in cans. The problem with this system is that in a hectic modern lifestyle, many people do not have the time to refill bottles or return cans to supermarkets, and new approaches may need to be considered.

The strengths and weaknesses of the system:

In Germany, the shift of responsibility and cost of recycling from the public sector to private industry, has been enormously beneficial. According to DSD statistics, 17,000 jobs have been created, primary packaging consumption has been reduced by 14 per cent and secondary packaging by almost 80 per cent.

There are, however, some drawbacks. Foreign companies have expressed concern that the laws create a trade barrier. Although foreign products are not required to carry the Green Dot, they still have to take responsibility for their packaging. Companies that ship their products long distances to Germany could end up paying the transportation costs of shipping the packaging back to the country of origin. Added to this, many manufacturers exporting to Germany claim that the domestic demand for the Green Dot label places imported goods at a disadvantage. Distributors and retailers are also deterred as without the Green Dot they share the responsibility of disposing of packaging.

These problems may find a solution as more countries follow suit, and a global standard of producer responsibility is reached. As an overall system, Green Dot has proven its worth, and is now licensed to 20 other EU nations that are striving to comply with the European Packaging Directive.

Different local authorities around Germany have different types of recycling banks. The picture below left shows the blue and yellow bins: blue is for plastic, aluminium and tin, and the yellow bin is for all cardboard and paper packaging.

Case study 7

Cuba: No waste

Cuba

In Cuba many people simply cannot afford to treat objects as disposable. Household goods, plastic, metal and glass containers, sheets of scrap metal and other items that in richer countries would be dumped, discarded and sealed in landfill limbo are reused, recycled and reincarnated in ingenious ways by Cuban people. Everything is valued for its inherent or material quality and reused and recycled according to this rule.

Kitchenware and utensils are fashioned by melting and mixing plastics from various sources; cups are made from salvaged soda cans and plastic containers; old food cans are turned into kerosene lanterns, containers and stoves.

Components from old and broken washing machines, televisions and modes of transport are valued because they can be used as replacement parts or reconstituted as elements of new appliances; a telephone becomes an electric fan; a computer monitor part of a television.

The Cuban economy struggled after the collapse of the Soviet Union. Prior to this Cuba had received economic aid from the communist state. In the 1990s the import and export of consumer goods and essentials was limited by both international sanctions and internal government enforced 'Post-Soviet Austerity Measures'. Whilst there has been considerable growth in the Cuban tourism and oil industries since then, the country still owes billions of dollars to creditors including the Paris Club and Russia. Whilst Cuba is not 'poor' by Latin American standards, Cuban citizens are still by no means wealthy.

Opposite top: Kerosene lamp.
Opposite bottom: Recycled kitchenware made by mixing plastics from various sources.

Ernesto Oroza, Nelson Rossel and Fabian Martinez formed the Laboratorio de Creacion Maldeojo in 1994. They began travelling the country photographing these recycled objects so that temporary as many of them are, they would be recorded for posterity. Oroza states that these objects are testament to the spirit of the Cuban people and their ingenuity: "The objects of necessity represent the world I live in, and they express our desire to invent and not let ourselves be overwhelmed by our problems… they reveal the hardships we've endured all these years and our hope that the state of things in Cuba will change. They are provisional objects that will disappear, but still deserve to be recorded."

These photos first appeared in *No Waste Pentagram Papers* no. 32.

Clockwise from top: A postcard used to hide wear on a light switch. A taxi sign made from a plastic bottle with a light inside for illumination. A spoon fence made with discarded aluminium sheets from an eating utensils factory. Sunglassses made from transparent plastic sheets and aluminium wire.

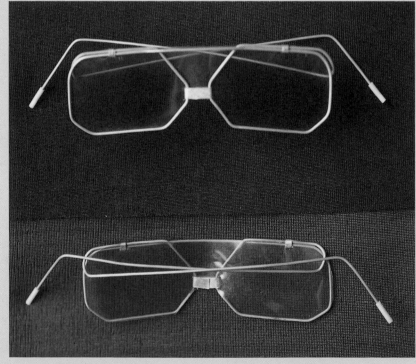

Clockwise from right:
A lamp made from a glass goblet,
toy marbles and pieces of acrylic.
Cups made of salvaged soda
cans and plastic containers.
Homemade guillotine mousetrap.
Kerosene stove made from tin
container and iron segments.
A homemade cooker igniter.

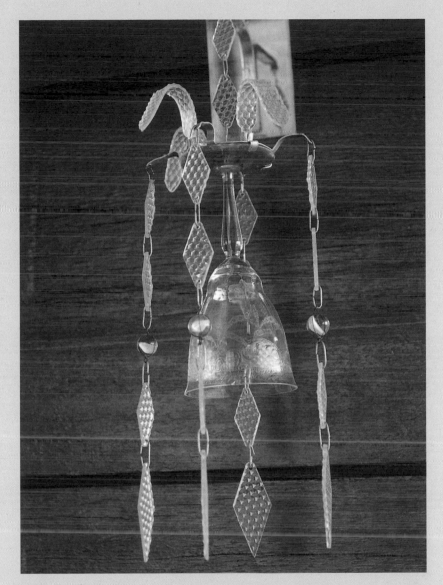

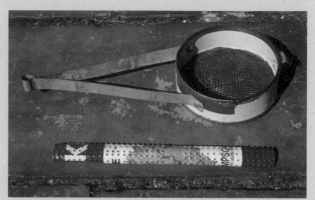

Clockwise from top: Recycled glass ashtray.
Colander made of salvaged metal.
Slingshot made of PVC piping and latex gloves.
Bicycle horn made of squeaky rubber toy.

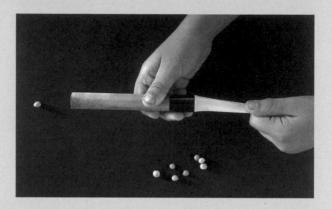

Clockwise from top:
Homemade aluminium containers.
Self contructed mini-bus made of
salvaged industrial materials.
Direct immersion water heater made
from salvaged industrial materials.
Toy oxen made of old bottles.
Pencil razor – used by soldiers.
Homemade pedal propelled vehicle.
Formula 1 toy cars made of ink and glue bottles.

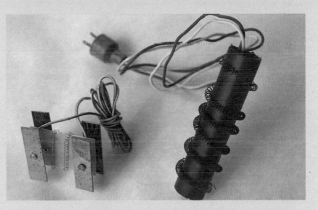

WANTED
For VICTORY
Waste Paper
Old Rags
Scrap Metals
Old Rubber
GET IN THE SCRAP
SELL TO A COLLECTOR OR GIVE TO A CHARITY

I am myself and what is around me, and if I do not save it, it shall not save me.

José Ortega y Gasset, Meditations in Quixote

Plastic

What is it?

The evolution of plastics has been a lengthy process and can be traced back to the 1830s when experiments with natural materials (rubber) and chemicals (sulphur), created a new material, vulcanised rubber. In the later part of the nineteenth century, more experiments led to lessening the natural product, and increasing the chemical element, to create celluloid and rayon. The invention of Bakelite, an entirely man-made material, occurred in the first decade of the twentieth century and can be considered the first true plastic. After that, the development of the material progressed in leaps and bounds, really taking off in the 1950s, with dozens of new types of plastics being developed for every possible use. Since then our consumption of plastics has increased by 2,000 per cent and continues to grow by about four per cent every year.

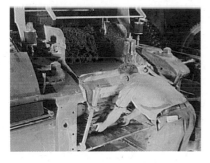

Top: Charles Goodyear, the inventor of vulcanised rubber.
Above: Early plastic production.
Right: The vast variety of plastic types we use today in our daily lives.

Why are they so popular?

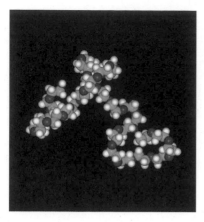

Above: The molecular structure of plastic.
Below: Plastics have come to replace glass as the most ubiquitous form of packaging.

Plastics are synthetic materials, made of petroleum, whose structures are based on the chemistry of carbon. These structures are by nature flexible. By changing components of the structure, one can control the properties of the final product. This makes it one of the most versatile materials available today. Added to this it is durable, water resistant, lightweight, inexpensive to produce and has excellent insulation properties. But all these advantages come at a price, and the environmental impact of plastics production is significant.

Above: The environmental impacts of plastics are far reaching, ducks swim in a river polluted with plastic bottles and bags.

The drawbacks

Plastics manufacture uses eight per cent of the world's oil, and produces harmful solid wastes as well as carbon dioxide, nitrogen oxide and sulphur dioxide emissions.

The chemicals used to stabilise, colour or impact on the properties of plastics have not been tested on humans, and their effect on human health remains unclear. Phthalates, brominated flame retardants, bisephonol A and dioxins are just a few of the chemicals which have raised particular concern. Many of them are thought to work as hormone disrupters and are possibly carcinogenic. Phthalates, for example, are hazardous toxins often used as a plastic softener in PVC, which may be released when the plastic comes into contact with saliva. PVC is a common plastic in use for baby toys.

Finally, it appears that plastics are virtually non-biodegradable. Their long polymer molecules are too large and too tightly bonded together to be broken apart and assimilated by decomposer organisms, and this means that they take a very long time to break down. It is still unclear what the exact time frame is, as they have not been around for long enough for this to be identified, but it is estimated that we could be talking about many hundreds of years. The disposable nature of plastics, and the alarming rate at which we consume them, makes landfill space a worrying problem.

The alternatives to landfilling are recycling or incineration. There have been recycling critics in the past that have claimed that due to the high calorific value of plastic, the best route is incineration for energy recovery. However, incineration of plastics, as an oil based substance, pollutes the atmosphere with carbon dioxide, contributing to climate change. The chemicals used in its production are then released again in its incineration, and the resulting ash has toxic qualities. Also, by destroying the products, new products have to be made, using even more resources. Finally, the energy produced by incineration is not renewable, and does not amount to the energy that can be saved by recycling.

The World Wide Fund for Nature (WWF) released a study which shows that bottled water sells for up to 1,000 times the price of tap water, but that the quality is often no better.

In 50% of cases the only difference is that bottled water has added minerals and salts, which does not actually mean the water is healthier. Furthermore, in the United States and Europe, there are more standards regulating tap water quality than that of bottled water.

In countries where tap water is unsafe, the study concludes that it is cheaper and just as safe to boil or filter water as it is to buy it in bottles.

Recycling Plastics

There are seven types of recyclable plastic:

1. PET — Polythylene terephthalate
Used for: bottled soft drinks and some bottled water, cooking oil bottles, oven-ready meal trays.

2. HDPE — Hi density polyethylene
Used for: plastic milk bottles, washing up liquid.

3. PVC — Polyvinyl Chloride
Used for: plastic pipes, outdoor furniture, bottled water, shrink wrap.

4. LDPE — Low Density Polyethylene
Used for: dry cleaning bags, carrier bags, bin liners.

5. PP — Polyproylene
Used for: bottle caps, margarine tubs.

6. PS — Polystyrene
Used for: foam meat trays, plastic tableware, vending cups, protective packaging.

7. OTHER
This category includes any recyclable plastic that does not fall into the above categories.

There are a number of problems at the moment with recycling plastic. The main one is that the various polymer types must be recycled separately, and as yet, there is no way of doing this mechanically. Although new technology is being investigated, most sorting of plastics is currently done manually into type and colour, using expensive and time-consuming human labour. Many products, including margarine tubs and rigid food containers, comprise many different types and colours of plastic, and are very difficult to recycle. In these cases, the energy and resources used in the recycling process may be more than that required to produce new plastic. One also has to bear in mind that plastic's main appeal is how cheap it is to produce, and the recycled plastic industry has to compete with these low costs.

Efficient recycling schemes have been set up for plastic bottles made of PET, HDPE and PVC, with a particular emphasis on PET. PET (known in some places as PETE) is a strong, lightweight plastic, able to withstand the massive pressure that carbonated drinks can generate inside a bottle (sometimes reaching six bar). PET is 100 per cent recyclable, and can be used to produce a multitude of products. About three quarters of reclaimed PET is used to make products such as fibres for carpets, fibrefill, apparel and geotextiles. Much of the remainder is extruded into sheet for thermo-forming, stretch blow-moulded into non-food containers, or compounded for moulding applications.

PET Plastic Bottle Sales, Recycling and Wasting in Britain, 1995 – 2002

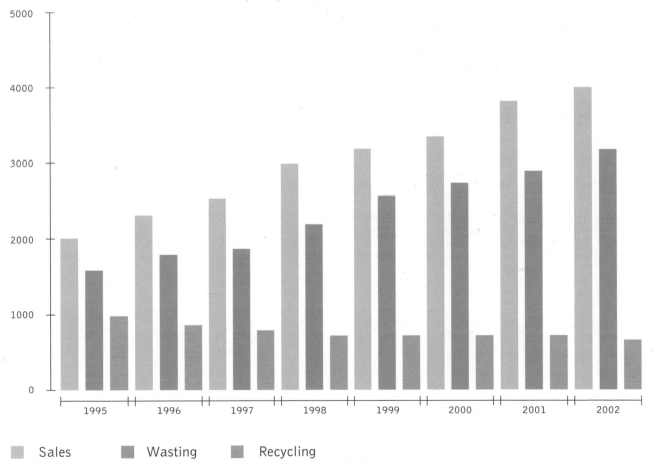

Sales   Wasting   Recycling

The recycling process of plastic

2
They are then washed and chopped into a more manageable size, and sorted once again using a floatation process (some will float and some will sink).

1
The different plastics are sorted according to their polymer type and colour.

3
From there, the plastic pieces are dried and melted. The melted plastic is filtered to remove contaminates.

4
The plastic is then squeezed out in strands, which are either chopped into little pellets...

5
... or spun into a fine fibre to make things like fleece or fibrefill.

What can I do?

As it is very difficult to recycle many types of plastic, the first step is to use less of it.

- Take a canvas shopping bag to the supermarket, or reuse the bags you were given last time.

- Be imaginative: turn yogurt pots into plantpots or use them for storage. Turn the top parts of soft drink bottles into cloches for plants.

- Try and buy food products with little or no packaging. Buy vegetables loose!

- Avoid disposable items like plastic cups and plates.

- Use refillable toner cartridges.

This page: When at the supermarket, be aware of unecessary plastic bags and packaging. Opposite: These tables by Yemm and Hart are made out of recycled plastic. (See p. 215 for details)

Recycle what you can. Find out from your municipality what plastics they accept. PET recycling schemes are now fairly widespread. Look for the symbol on the bottle/package to check whether it is recyclable.

- Make sure to take the bottle tops off – these are made of a different polymer and must be recycled separately!

- Buy recycled products such as certain brands of bin bags, and try buying products in refillable packaging such as those offered by the Body Shop and Ecover.

- Suggest to your local authority that they buy street furniture made of recycled plastics.

- In the office, you may be able to join Save-a-Cup, or a similar scheme, which collects a specific plastic product on a national scale.

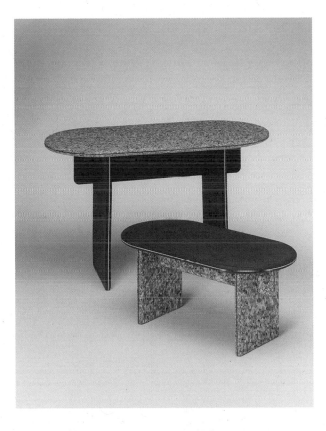

↻ Some plastic recycling facts:

Only 7% of plastic in Britain and 5% in America is recycled.

Plastic makes up 11% of household waste, 40% of which is plastic bottles.

60% of litter on beaches is plastic.

One study reported that 1.8 tons of oil are saved for every ton of recycled PET produced.

The future

Due to the problematic nature of plastics, alternatives are constantly being sought out. Biodegradable plastics, or biopolymers, have been looked into extensively. These are plastics mixed with starch or bacteria that can break down like any other organic matter. However, the technology has not been perfected. Some of these plastics will only break down when exposed to sunlight, and often they are buried so deep in a landfill that no sunlight falls upon them, meaning that in essence, they are still non-biodegradable. They are also very expensive to produce (on average ten times more expensive than traditional plastics), and so lose out on one of plastic's main appeals. Finally, there is a concern that biodegradable plastics will seriously impede the future of plastics recycling (by getting confused in the waste stream), without providing a long-term, viable alternative.

Feedstock recycling is another technology currently being explored. This is a type of plastic recovery, which breaks down polymers into their constituent monomers, which can be used again in refineries or petrochemical production. Feedstock recycling is more tolerant to impurities than mechanical recycling, and if it could be implemented on a large scale, could be economically viable.

An incentive for further research into plastic alternatives is the burgeoning price of oil, which is pushing up production costs. Scientists are looking at carbon based non-petroleum materials, which may become more cost-effective and a more attractive option in a world less reliant on fossil fuels.

An incentive for further research into plastic alternatives is the burgeoning price of oil, which is pushing up production costs.

Coloured PET, such as bottles used for some soft drinks, is known as 'Jazz PET', and has a market value just a fraction of that of clear PET.

This is because when it is recycled it creates a black fibre, rather than a white fibre, which is much less desirable. This can be used to produce street furniture, car bumpers and drainpipes, but ultimately its uses are limited.

↻ Biopolymers

Biopolymers are another name for biodegradable plastics. They can be divided into four main categories: starch, sugar, cellulose and synthetic materials. Each category has its own advantages and disadvantages.

Starch: This is a natural polymer that can be derived from potatoes, maize, wheat, tapioca and other sources. The chemical constitution of the starch can be modified so that it can be melted and reformed. The drawback is that starch can only sustain brief contact with water before it starts to disintegrate, and is therefore not suitable for packaging liquid, but more suited to plastic forming processes such as injection moulding.

Sugar: Sugar based polymers, or polyalctides, are derived from sucrose or starch and then undergo a process of bacterial fermentation. This means that bacteria actually grow granules of a natural plastic within their cells, and this plastic is then harvested. By varying the nutrient composition of the bacteria, it is possible to tune the properties of the material (therefore sugar based plastics can be water resistant). This is a developing field, although the high price of polyalctides means that it's not quite ready to be used for packaging.

Cellulose: Cellulose, the main ingredient in cellophane – used for packaging of sweets and cigarettes – is biodegradable and can be composted. It's got a long history of use both as packaging and for other materials such as film. However, it's fast falling out of favour, as like the sugar based plastics, the price is high, and it can't compete with other polymers such as polypropylene.

Synthetic: It is possible to use synthetic compounds to create biodegradable plastic. There are numerous experiments with these new compounds, and though the price remains high, there are high hopes for their development. It is important to bear in mind that although these plastics will be biodegradable, their primary ingredient will still be petroleum.

The future of biopolymers appears to lie in hybrid combinations between the cheap, starch based plastics and the bacteria based, more expensive polymers.

UK

Recoup
*Recycling of used
plastic containers.*
1 Metro Centre
Wellbeck Way
Woodston
Peterborough, PE2 7UH
Tel: 01733 390 021
www.recoup.org

British Plastics Federation
*Information about plastics
manufacturing and recycling.*
6 Bath Place
Rivington Street
London, EC2 3JE
0207 457 5000
www.bpf.co.uk

Save a Cup
*Collects hard-wall polystyrene
cups covering the vending, food
services and plastic industries.*
Suite 2, Bridge House
Bridge Street
High Wycombe, HP11 2EL
Tel: 01494 510 167
www.saveacup.co.uk

UNITED STATES

American Plastics Council (APC)
*All about plastics, the future
of plastics and future products
made of plastics. With links to
other sites run by the APC.*
1801 K Street, NW, Suite 800
Washington DC, 20036
Tel: 202 862 5100
www.americanplasticscouncil.org

Association of Postconsumer Plastic Recyclers
*A website geared towards
industry professionals, but
with interesting articles
and information about the
problems with plastics usage.*
1300 Wilson Boulevard
Arlington, VA 22209
Tel: 703 741 5578
www.plasticsrecycling.org/

National Association for PET Container Resources (NAPCOR)
*Information about production
and recycling of PET.*
PO Box 1327
Sonoma, CA 95476
Tel: 707 996 4207
www.napcor.com

Plastic Loosefill Council
*For recycling those
Styrofoam peanuts that
come in packaging.*
PO Box 21040
Oakland, CA 94620
Tel: 510 654 0756
www.loosefillpackaging.com

CANADA

Canadian Plastics Industry Association
5915 Airport Road, Suite 712
Mississauga
ON, L4V 1T1
Tel: 905 678 7748
www.cpia.ca

Canadian Polystyrene Recycling Association
7595 Tranmere Drive
Mississauga, ON L5S 1L4
Tel: 905 612 8024
www.cpra-canada.com

AUSTRALIA

Plastic and Chemicals Industry Association (PACIA)
Click on the recycling icon for more information about plastics recycling.
Level 2, 263 Mary Street
Richmond, Victoria 3121
Tel: 61 3 9429 0670
www.pacia.org.au

SOUTH AFRICA

Plastics Federation of South Africa
Private Bag X68
Halfway House, 1685
South Africa
Tel: 27 11 314 4021
www.plasfed.co.za

NEW ZEALAND

Plastics New Zealand
Easy to use, comprehensive resource on plastic and recycling.
PO Box 76 378
Manukau City
Tel: 64 9 262 3773
www.plastics.org.nz

EUROPE

Petcore
The PET recycling association of Europe.
Av. de Cortenbergh 66 box 5
1000 Brussels
Belgium
Tel: 32 2 736 72 64
www.petcore.org

Case study 8

Australia

Coles Bay, Tasmania and the plastic bag conundrum

Plastic bags are bad, bad news. Like all plastics, they are hazardous to produce and take hundreds of years to decompose. But what's really worrying is the casual way in which they are dispensed. Every time we go to a shop, we are offered plastic bags, and we take them willingly. It is thought that an average Briton takes home an average of 35 plastic bags a week. Some of these bags will be reused once – for rubbish or instead of clingfilm – but virtually all of them find their way to the bin and then to landfill.

Apart from the problems regarding their production, plastic bags have numerous detrimental environmental impacts. It is thought that up to 100,000 animals, including birds, whales, seals and turtles die every year by ingesting or getting tangled in plastic bags. The litter caused by plastic bags can cause blocked drains and flooding. And to top it all off they cost retailers hundreds of millions of pounds – a cost that is passed on to the consumers.

So what's being done about it? Different countries the world over are taking different approaches to the plastic bag conundrum. In Bangladesh, a ban was put on production and distribution of plastic bags, which had blocked up city drains, causing floods that killed dozens of people. In Ireland a tax has been imposed on all shopping bags, which, according to recent reports has led to a 90 per cent reduction in plastic bag usage (see Case study 9, p. 136). In Italy, a similar tax has been levied. In Canada, all big supermarket chains must accept bags for recycling, leading to a 45 per cent recovery rate of plastic bags. But for the most part, the reaction has been not to do anything. Whilst governments in Australia, the United States and Britain discourage the use of plastic bags by way of public information campaigns and provision of alternative carrier bags, no actual laws have been passed as yet to put a stop to our escalating usage.

Left: Burning plastic bags releasing toxins into the atmosphere.
Below: A leaflet produced by PlanetArk warning of the dangers that plastic bags pose to wildlife. Here a Brydes Whale has died after ingesting six metres of plastic.

A gutful of plastic

Dying whales may move into shallow water to avoid drowning.

Floating rubbish, such as plastic, can be easily swallowed, and since it cannot be digested or passed, it stays in the whale. Plastic in the gut can prevent the animal from digesting its food and may lead to death.

In August 2000, an 8 metre Bryde s (pronounced broodas) whale stranded close to central Cairns in north Queensland. It died soon after.

An autopsy found that the whale s stomach was tightly packed with plastic - almost 6 square metres of it! The whale had swallowed supermarket bags, food packaging, three large sheets of plastic 2 metres long and fragments of garbage bags.

Bryde s whales feed by swallowing large amounts of water. They use baleen, the fringe along the tops of their mouths, to sieve out small fish and other food.

Please be careful with your rubbish.
A plastic bag dropped in the street washes into stormwater drains which empty into the ocean. If you see plastic in the street, don t let it become whale food - pick it up!

Let's keep plastic off the street
and out of the ocean!

In certain towns, specifically in Australia, residents have taken matters into their own hands. The winner of the Australia Day Local Hero Award 2005 was a man named Ben Kearney, the owner of the local bakery in a small Tasmanian town called Coles Bay. With a population of a mere 175, Coles Bay is a fairly major tourist destination, as it borders the Freycinet National Park, and is a migration point for whales. Every year, the residents found that they were consuming 350,000 check-out bags. The environmental group, PlanetArk, worked alongside Kearney to make Coles Bay Autralia's first plastic bag free town in 2003. Every household in the town has been given five large calico bags for carrying shopping, and the bakery seals its loaves in a biodegradable plastic made of tapioca starch. Tourists must buy a calico bag for AUS$2.50 (£1, $1.80) or a reusable paper bag with a handle for 25 cents (10 pence, 19 US cents). By all accounts, the impact has been enormous, and the reaction of residents, visitors and the media has been overwhelmingly positive. So much so, in fact, that other towns in Australia, namely Kangaroo Valley, Huskisson, Mogo and Oyster Bay, have followed suit.

Coles Bay is now putting in motion further moves to increase sustainability. Water usage restrictions are being put in place, and moves are being made to create a more efficient recycling system. The town has applied for environmental funding and recognition usually saved for businesses. The Australian Government has also taken heed, and the Australian Retailers Association is looking at ways of phasing out plastic check-out bags throughout Australia by 2010.

Top: A view of Coles Bay. Photographer Kip Nunn. www.tasmanianphotography.com.au
Right: Ben Kearney holds up one of the reusable bags now being used in his bakery.

How to ban plastic check-out bags in your town[10]

1. Make sure your town or suburb is capable of going plastic check-out bag free. A town or suburb that is capable of going plastic check-out bag free has all locally owned supermarkets and retail outlets. This makes it easier as the decision to ban plastic bags can be community driven at the very local level.

2. Designate a local champion to coordinate your town's plastic bag free campaign.

3. Talk to local community groups such as rotary clubs that may be able to help you.

4. Compile a list of retailers in your town that use plastic check-out bags. Once some of the shops start going plastic bag free, use those names when you are persuading other retailers to stop using plastic bags.

5. Order paper, calico and polypropylene bags for every retailer to sell. www.planetark.com/bagorders has links and details of reusable bag options, prices and ordering details.

6. Write to all retailers in your town to invite them to join your plastic bag free town campaign.

7. Run an information night for retailers and the local community to attend. Make sure you have researched the benefits of banning plastic bags, so you can answer any questions you may be asked.

8. Get retailers to sign off on their commitment to go plastic bag free.

9. Decide on a design to go on your town bags: by having one design for the whole town, you can save money as there will be only one set up cost, this also ties the community effort together.

10. Set a launch date for when your town will be plastic check-out bag free. This will be when the reusable bag orders arrive and can be distributed throughout the town.

11. Make a local launch date announcement, so that all the residents of the town are aware that from now on there will be no plastic check-out bags. Contact the local media to cover this.

12. Arrange a 'plastic bag free town' launch day. Invite the local mayor to speak, and make sure there is media coverage.

10 From www.planetark.com

Case study 9

Ireland

It's in the bag — Ireland's plastic bag tax

In March 2002, Ireland became the first country to introduce a plastic bag tax, known as 'plas-tax', of 15 cents (10 pence, 18 US cents) per bag. This was no stealth tax to save the government from economic disaster, but a carefully calculated scheme to change consumer behaviour, to raise awareness and to reduce landfill and litter problems resulting from excessive plastic bag usage.

Up until the introduction of the tax, Irish shoppers were using 342 bags per year each — a total national consumption of 1.2 billion bags. Retailers were spending 50 million euros on supplying these bags, and as ever, the environmental implications were not exactly rosy. Since introducing the tax, Ireland has seen a 90 per cent drop in consumption: that's approximately 1 billion fewer bags consumed annually.

Heavier weight, reusable bags are exempt from the tax, as are bags used for meat and fish, bags for unpackaged produce or other foods without packaging. Reusable shopping bags are taking the place of plastic disposables, and some of the plastic bag manufacturers have benefited by picking up on this market.

The Environment Minister, Martin Cullen said that the improvement had been "immediate and plain to see". In the first six months alone 3.5 million euros were raised in revenue, which he assured would be spent on other environmental projects. He added: "It is clear that the levy has not only changed consumer behaviour in relation to disposable plastic bags, it has also raised national consciousness about the role each one of us can and must play if we are to tackle collectively the problems of litter and waste management."[11]

England, Australia and New York City are now considering implementing a similar tax.

11 BBC News, "Irish Bag Tax Hailed a Success", news.bbc.co.uk/1/hi/world/europe/2205419.stm

Case study 10

Barnet compulsory recycling scheme

UK

Switzerland is not the only one to take a punitive approach to recycling. In a scheme launched in 2004, Barnet, a north London borough, has instituted a £1,000 ($1,750) fine if residents fail to participate in their black box scheme. The scheme only applies to houses, not flats, and there is no set amount of recycling a household has to undertake to meet the terms of the compulsory scheme. Only those "deliberately and persistently" failing to put anything in their black recycling box face further action.

The scheme was implemented as follows: a pilot scheme of 25,000 houses was run for a one year period. Each of those houses were sent a letter informing them of the council's intentions. Once the scheme kicked in, Street Enforcement Officers monitored the participating properties. Over the period of the pilot scheme, recycling statistics improved by 18 per cent, and this was deemed sufficiently successful to implement a borough-wide scheme, which is now in place, covering 113,000 households.

Households who are not compliant with the scheme are sent a warning letter. If they continue not to comply, one of three 'Recycling Assistants' employed by the borough pays a home visit, and warns of the possible repercussions. Those who persistently refuse to recycle will face prosecution. So far, five such households have been identified, and the authorities are now taking this further. They will have to prove that glass, paper and cans are being put in the refuse bin instead of recycling, and will then have a legal standpoint from which to fine the households, under section 46 of the Environmental Protection Act 1990. They have already issued a legal notice on one such household in Totteridge.

The scheme is being monitored by DEFRA, The Department for Environment Food and Rural Affairs, and is to be considered for implementation elsewhere in Britain.

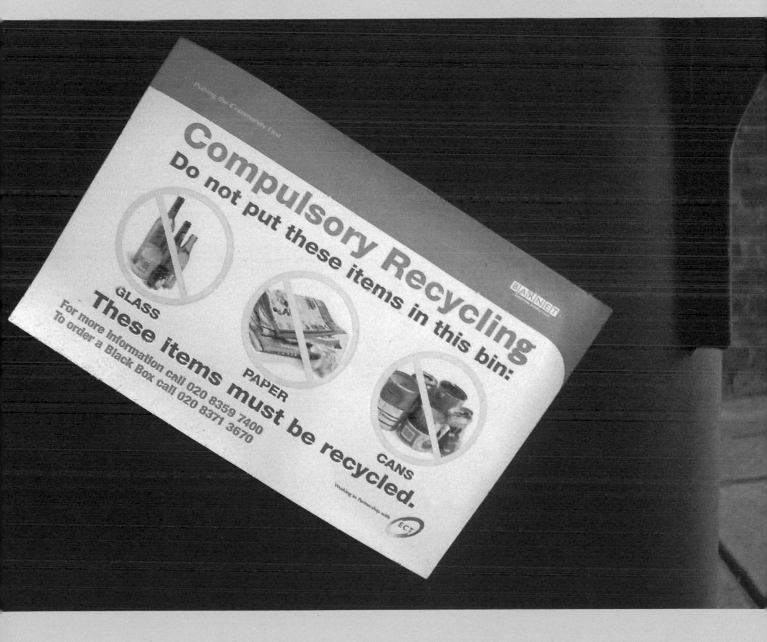

Earlier periods in human history were marked by the material that distinguished the era — the Stone Age and the Bronze Age for example. Our age is simply the Material Age, an age of excess, whose distinguishing feature is not the use of any particular material but the sheer volume of materials consumed.

Lester R Brown, Eco Economy

Household and office waste

As fashions and fads change ever more quickly, expensive items that once would have lasted many tens of years are more casually disposed of.

We live in a disposable culture. Planned obsolescence and throw-away products make economical sense – creating new jobs and promoting economic growth. Ecologically, they make no sense at all. Each day we use hundreds of products – from mobile phones to ballpoint pens. Some of these have a lifespan of a single day, others a few years, but eventually they all become waste. Many of these products are composites of numerous materials: plastic and metal and rubber and glass. A mobile phone, for example, contains 500 to 1,000 parts. It is often impossible to break these products down to the miniscule components and recycle each material individually.

This section will take a broad approach to some of the household and office items you use regularly which can be disposed of in a responsible way, causing a minimum of environmental damage.

E-WASTE

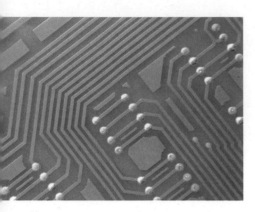

Nothing exemplifies disposable culture more than electrical goods. Over the last two decades electronics have become an integral part of our daily lives. Televisions, computers, mobile phones and MP3 players are just some of the electronics that we have come to rely on. As product innovations multiply to keep up with our evolving demands, the lifespan of electronics has shrunk, equipment often becoming obsolete as soon as faster, more sophisticated merchandise hits the shelves. Discarded electronic equipment is growing three times faster than any other municipal waste category, and is known as e-waste. It now makes up five per cent of all municipal solid waste worldwide – nearly the same amount as all plastic packaging.

The main component of e-waste is white goods (large household appliances such as fridges, cookers and washing machines). This is fast followed by IT equipment and then by televisions. These products often contain hazardous and toxic chemicals that can damage the environment if they are incinerated or landfilled. Televisions and computer monitors use cathode ray tubes, which contain lead; older electronics contain mercury in their relays and switches; and fridges contain powerful greenhouse gasses called chlorofluorocarbons (CFCs).

Reuse and recycling prevents electronic items from reaching landfills and leaching these chemicals into the earth. Reuse should always be the first choice, as it saves much more energy and uses no virgin resources. There are numerous schemes that extend the lifespan of discarded electric products and sell or donate them on to other users. These should always be investigated before recycling.

Reuse and recycling prevents electronic items from reaching landfills and leaching these chemicals into the earth.

A stack of white goods, all containing toxic chemicals that can leach into the soil.

Computers and IT Equipment

Even if a computer does not work, or is not of sufficient standard to be reused, there may still be some components that can be salvaged and reused. Floppy disks, cords, speakers and keyboards are often fairly standardised, and can be removed from the computer body and reincorporated into other computers.

If there really is nothing that can be salvaged, as of August 2005, the Waste Electrical and Electronic Equipment Directive (WEEE) has required all producers in the EU to take responsibility for their discarded products. This means you can return your equipment to the manufacturer when you have finished with it, and they are legally obliged to dispose of it safely. If you are not based within the EU, or want to be certain that your computer is not exported, and unethically recycled, there are schemes that will dismantle the computers and recycle each material individually (plastic, glass, copper, steel etc.). Since this will almost certainly require a high degree of manual labour, there may be a charge for this service. Generally speaking, the cost to recycle computers ranges from £4.50–£6 ($8–$20), and they contain about £1.70 ($3) worth of recyclable precious metals, most of which are found in the circuit board and wiring.

Below: Toner cartridges ready for recycling – most offices can run a toner cartridge recycling machine if they cannot obtain refillable cartridges.

- Use refillable printer cartridges. If you have an inkjet computer, refilling them is straightforward either on a DIY basis or by sending them away for refilling.

- Always remember to clean off personal information on any IT equipment before passing it on!

Computer facts

The average lifespan of computers in developed countries has dropped from six years in 1997 to just two years in 2005.

183 million computers were sold worldwide in 2004 – 11.6% more than in 2003.

By 2010, 46% of people in Mexico will own a computer.

The cathode ray tubes in monitors sold worldwide in 2002 contain approximately 10,000 tons of lead. Exposure to lead can cause intellectual impairment in children and can damage the nervous, blood and reproductive systems in adults.

Cadmium, used in rechargeable computer batteries and mobile phones can bio-accumulate in the environment and is highly toxic, primarily affecting the kidneys and bones.

Mercury, used in lighting devices for flat screen displays can damage the brain and central nervous system, particularly during early development.

Brominated flame retardants, used in circuit boards and plastic casings can build up in the environment. Long-term exposure can lead to impaired learning and memory functions and can also interfere with thyroid and oestrogen hormone systems.[12]

12 Statistics and information available from Greenpeace information sheet, "Toxic Tech", *Greenpeace International*, May 2005

White Goods

The category of white goods contains within it refrigerators, washing machines, dishwashers and other large domestic appliances. There is a much stronger tradition of recycling white goods, due to regulations requiring the recovery of CFCs prior to disposal. Prior to 1994 almost all appliances used CFCs as both refrigerant and as foam blowing agent. After 1994 these were mainly replaced with HFC as refrigerant and HCFC foam blowing agents. These are an improvement, in that they don't deplete the ozone layer, but they are also potent greenhouse gases. In fact, left unchecked, by 2050, HFCs will do as much damage to the climate as the traffic fumes of all the world's passenger cars. In the view of Greenpeace, it was a case of "out of the frying pan into the fire". As a result, the organisation helped develop an alternative technology, which avoids the use of HFCs altogether. They named it Greenfreeze, and this eco-friendly fridge is now available through all the big manufacturers.

As with computers, when disposing of white goods, first priority should be given to refurbishment and reuse, rather than recycling. Charities often run a collection and recycling scheme for white goods. See if there's one in your area. If the appliance is broken or below usable standard, arrange for safe disposal via recycling. The recycling process usually involves removing hazardous materials, and then shredding and recycling the steel and plastic.

- Make sure your fridge is empty and cleaned before it is sent to donation or recycling.

- If possible, take off the fridge door, so that kids playing around don't get caught inside.

- If local traders offer to dispose of your fridge for you, contact your local Environmental Agency to make sure they intend to dispose of it responsibly.

A fridge donation facility in Britain.

Mobile Phones

Mobile phones are one of the most quickly discarded items of consumer electronics. Service providers insist on upgrading their customers annually, leaving an average of 15 million handsets discarded each year in the UK alone. The rechargeable batteries on these phones and other components such as the LCD display contain toxic chemicals including cadmium, rhodium and lead. As with other electronic goods, if the handsets end up in landfills, these chemicals will eventually leach into the soil. It is said that one battery of the older, larger mobiles (of the 'brick' variety) is enough to pollute 600,000 litres of water with cadmium.

Recycling schemes for mobile phones are becoming more widespread. Oxfam, ActionAid and other NGOs refurbish old mobiles and pass them onto developing countries. Most phone shops will take the handsets back, and pass them on to charities or to the manufacturers, who may reuse or recover useful parts such as aerials, battery connectors and speakers, and send the dangerous items (primarily the batteries) to specialist reprocessing plants to extract the heavy metals.

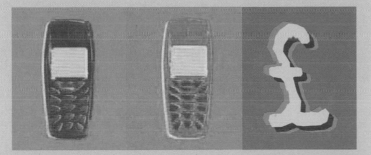

Christian Aid is one of many organisations that can arrange for the reuse/recycling of your old handset.

Batteries

Although waste batteries constitute a small percentage of the solid waste stream, they are a serious environmental concern due to their high concentration of heavy metals. Lead, cadmium, nickel hydride and lithium are all found in varying quantities in most types of batteries. If these are landfilled they can contribute to soil and water pollution. Cadmium, for example, can be toxic to aquatic invertebrates and can bio-accumulate in fish, disrupting food chains and ecosystems. If cadmium or lead are ingested, they are known to promote neurological, pulmonary, renal and genetic disorders. Until recently, mercury was also a concern, although since 1994, mercury has been phased out of most household battery production.

When discussing the recycling process for batteries, it is useful to differentiate between wet cell batteries – used in cars and other motor vehicles – and dry cell batteries, which constitute most household batteries.

Wet Cell

Car batteries have a high recycling rate (reaching 90 per cent in the UK). The market value of the extracted silver or lead is greater than the cost of recovering the spent battery waste, and this is a major incentive for recycling. This has been further enforced by legislation banning the casual disposal of large batteries, and so very few wet cell batteries ever reach landfills.

The recycling procedure is as follows:

- The batteries are crushed by a huge press, which breaks them into component parts.

- The plastic is washed, dried, granulated and recycled.

- The lead is melted down to make more car batteries as well as other lead products such as guttering and x-ray shields.

- The acid is treated and neutralised.

- The distilled water is purified and reused.

Dry Cell

Dry cell batteries are a more difficult prospect. Firstly it is less economically viable to recycle household batteries – it is thought to cost three times as much to recycle than to treat and dispose of batteries. Secondly, there are a few different types of household batteries, each of which need to be recycled in different ways, and this adds sorting costs to the procedure. As a result, there are fewer recycling plants – in fact, until recently there were NONE in the UK. All batteries sent for recycling had to be shipped over to France, incurring further costs. Less than two per cent of disposable batteries and five per cent of rechargeable batteries in the UK are recycled.

This situation is changing. New legislation is making battery recycling a priority, and battery banks are cropping up more regularly. Many kerbside schemes now accommodate battery recycling and certain retail outlets provide bins for battery disposal. It is hoped that using technology developed for the steel recycling market, it will become more viable to set up more battery recycling plants.

Above: Dry cell batteries.
Below: Wet cell batteries.

What you can do:

- Use rechargeable batteries. These have longer lifespans, and can generally be sent back to the manufacturer if there is no recycling scheme available in your area.

- Check what recycling schemes are available in your area and use them!

- If there does not appear to be any option available, speak to the shop that sold you the batteries, or to the battery manufacturers themselves and ask them to provide a solution.

Tyres

Tyres are made of vulcanised rubber – a form of chemically enhanced rubber that is designed to last. They are bulky, a potential fire hazard and contain environmentally toxic substances. In landfills they break down very slowly, and when incinerated, they produce large quantities of pollutants.

One solution to tyre disposal is retreading. This means either replacing the tread section or the entire outer surface of a tyre. Truck tyres can be retreaded up to three times, but due to the strict standards in place, car tyres can only be retreaded once. In the UK, 50 per cent of truck tyres and 100 per cent of aeroplane tyres are retreaded. However, very few domestic vehicle tyres are being retreaded, and the numbers are falling. Budget tyres are taking over the market that was supplied by retreaded tyres, and governments are being urged to support the retread market.

Tyres can be exported to developing countries to be retreaded and reused.

Women at the REFDAF project in Dakar using tyres as a building material. (See Case study 2, p. 57)

The alternative to retreading is recycling. There are numerous ways to recycle tyres:

- Tyres can be used whole as drainage systems for landfills (and are therefore exempt from landfill tax).

- Tyres can be granulated into crumbs and used in children's play areas, carpet underlay, rubber car mats and sports surfaces.

- They can be de-vulcanised and re-converted into rubber. This rubber is of diminished quality, and its uses are therefore limited, but it can be used in conjunction with virgin material to produce new tyres, conveyor belts and shoes.

- Tyres, like plastic, have a high calorific value. If this can be properly harnessed, tyre incineration can be used to create energy. A few technologies are being developed that would allow this energy to be released, without also releasing toxic fumes. These technologies are pyrolysis and incineration in cement kilns.

- Tyres can be exported to developing countries to be retreaded and reused.

What you can do:

- Treat your tyres well – avoid potholes, check pressures regularly, try not to brake heavily or take corners too sharply.

- When your tyres wear out, consider purchasing a retreaded tyre.

- Dispose of used tyres responsibly by getting in contact with your local tyre retailer. There may be a small charge for arranging recycling.

Textiles

Textiles are defined as items that are made from woven or knitted cloth. Textiles made of synthetic fabrics do not decompose, and clothes made from natural fabrics rot and produce methane. A fairly sophisticated recycling network is in place to deal with waste textiles.

Textiles can be divided into two categories; post-industrial waste – waste that arises during the manufacture of fabric and garments, and post-consumer waste – used clothes and other textiles.

Post-industrial waste and post-consumer waste that is unsuitable for resale/reuse are recycled in the same way. The fabric is sorted by type, colour and grade (heavy/light). The material is shredded into fibres. This material is known as shoddy. If the shoddy is of high grade, it is then mixed with virgin fibre and spun into new fibre for weaving or knitting. Low grade shoddy will be recycled as factory wipes, stuffing for cushions and pillows, carpet underlay, mattress padding and cotton swabs.

Post-consumer waste that is suitable for reuse can be recycled in charity shops, or placed in clothes recycling bins. The Salvation Army, Oxfam and the Red Cross are just a few of the charities that run recycling schemes. Some of them even have a door to door textile collection service. The clothes you donate will be given to the homeless, sold in charity shops to generate revenue for the charity, or sent to developing countries in Africa, Asia and Eastern Europe. If some of the clothes are unusable, the charities will sell them as raw materials to the textile recycling industry.

Charity shops have been around for many years as this image from Lynwood, California in the 1950s shows.

Recycling fibre:

- Reduces the need for landfill space.

- Saves energy, as materials do not have to be sourced and transported, and fibres do not need to be re-dyed or scoured.

- Reduces pollution that would result from the transportation and from the dying and fixing processes virgin materials require.

- Saves water – especially in the case of raw wool, where the washing procedures require large quantities of water.

What you can do:

- Take your old clothes to a charity shop or a recycling bank.

- Always tie shoes together – they are useless if they lose their pair!

- Keep clothes you are going to give to a charity shop in a dry place. Mouldy clothes cannot be re-sold.

- Similarly, do not leave clothes on the doorstep of a charity shop overnight. They may get dirty or rained on. Wait till the morning and take them inside.

- Buy second hand clothes.

- Buy cloth wipers made of recycled fabric instead of paper towels.

- Buy clothes made with recycled fibres.

Wood

Household wood waste comes from old furniture, fencing and DIY off cuts. Wood is a valuable resource. The harvesting and processing of virgin timber uses significant quantities of energy and water. In the UK, much of the timber used is softwood, 85 per cent of which is sourced in sustainable forests in Scandinavia and the Baltic States. However, approximately ten per cent of wood in Britain comes from tropical countries, and some of that may be illegally logged timber from rainforests, and other ancient forests.

- If buying new timber, look for the FSC (Forest Stewardship Council) logo. This will ensure that the wood is ethically and sustainably produced.

Above: This chair is being dismantled and repaired for resale.
Below: A rocking chair made of recycled wood, designed by David Meddings. (See p. 209 for details)

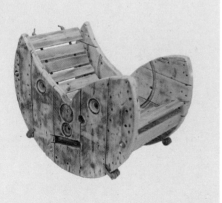

Wood is easily reused and recycled.

- If furniture is in good condition, a furniture reuse project may be able to collect it and pass it on to low income families.

- Architectural salvage groups will take items like floorboards, door frames, doors and staircases.

- If reuse is not an option, there are occasional facilities set up for recycling. Recycled wood can be used in mulch, pet bedding, chipboard production, paper production and can be used as fuel in timber fired boilers.

Household Hazardous Waste

Household Hazardous Waste (HHW) is a generic term for all domestic waste products with the characteristics of ignitability, corrosivity, or toxicity, and includes antifreeze, acetone, floor polish, paint, pesticides, oven cleaners and other solvents. Casually disposing of these materials can be dangerous on many levels: they can cause injuries to workers in waste management operations, can cause equipment and property damage by igniting in refuse collection trucks or transfer trailers, and, of course, they cause environmental contamination by leaching into the soil.

The solution to this is fairly complex. Firstly, there has to be some level of source control; lessening the quantity of hazardous waste that is being produced, and making people aware of alternatives. For example, using baking soda as a benign cleaning product, using cedar chips instead of mothballs and using soap based rather than chemical based cleaners. Ecover makes a broad range of ecologically friendly cleaning products, and even has facilities for bottle refill.

There is also potential for reuse of some HHW products, especially paints. There have been several success stories involving setting up paint reuse schemes in communities, allowing local projects or neglected community buildings access to free paint.

Finally there are some, admittedly limited, facilities for recycling.

- Recycling motor oil is the most common HHW management practice. It is collected at service stations, recycling centres and landfills, and can be easily recycled. It is boiled and filtered and can be re-refined into lubricating oil, using only one third of the energy that refining crude oil requires.

- Solvents can be separated and bulked for use as a supplementary fuel.

- Antifreeze can be processed and distilled and reused as antifreeze or used as a chemical agent in mineral and cement processing.

These schemes are fairly few and far between. At present it is best to consult your local environmental agency for advice about safe disposal of HHW.

UK

ICER
Industry Council for Electronic Equipment Recycling.
6 Bath Place
Rivington Street
London, EC2A 3JE
Tel: 0207 729 4766
www.icer.org.uk

Waste Care
A private waste collection company that will arrange for hazardous waste (including mobile phones and batteries) to be responsibly disposed of.
Richmond House
Garforth
Leeds, LS25 1NB
Tel: 0113 385 4321
www.wastecare.co.uk

Envirogreen
A private waste disposal company specialising in batteries and Household Hazardous Waste.
Regus House
268 Bath Road
Slough
Berkshire, SL1 4DX
Tel: 0845 712 5398
www.envirogreen.co.uk

Donate a PC
Noticeboard where you can donate your PC to charity.
www.donateapc.org.uk

Computer Aid
A non-profit supplier of recycled computers to developing countries.
Ground Floor
433 Holloway Road
London, N7 6LJ
Tel: 0207 281 0091
www.computeraid.org

Secure IT Disposals Ltd
Private company specialising in disposal and recycling of redundant computers.
104-108 Floodgate Street
Digbeth
Birmingham, B5 5SR
Tel: 0121 643 2996
www.sitd.co.uk

Rebat
An initiative for British battery recycling.
AEA Technology
Harwell Business Centre
Building 551 Didcot
Oxfordshire, OX11 0QJ
Tel: 01235 434 061
www.rebat.com

Actionaid, National Recycling Unit
Recycles used batteries and printer cartridges.
Unit 14 Kingsland Trading Estate
St Philips Road
Bristol, BS2 0JZ
Tel: 0117 304 2390
www.actionaidrecycling.org.uk

Recycle Appeal
Recycles printer cartridges and mobile phones for charity and puts you in touch with the various aid organisations that will be able to help.
Redeem plc.
Unit C–E Etna Road
Falkirk, FK2 9EG
Tel: 08712 50 50 50
www.recyclingappeal.com

Various charities, such as Oxfam, ActionAid and Scope will recycle mobile phones. Find their contact details in the NGO directory at the back of this book.

The Mobile Phone Recycling Company
Buys and sells used mobiles, as well as raising cash for charities through mobile phone donation.
Contact via website.
www.mobilephonerecycling.co.uk

Furniture Reuse Network (FRN)

48–54 West Street
St Philips
Bristol, BS2 0BL
Tel: 0117 954 3571
www.frn.org.uk

Salvo

*Directory of architectural
salvage resources in the UK.*
www.salvoweb.com

Recycle Wood

*Information and links
about wood recycling.*
Run by WRAP
Tel: 0808 100 2040
www.recyclewood.org.uk

Timber Recycling Information Centre

*User friendly website with
information and contacts
for wood recycling.*
Chiltern House
Stocking Lane
Hughenden Valley
High Wychombe
Bucks, HP14 4ND
Tel: 01494 569600
www.recycle-it.org

Textile Recycling Association

PO Box 965
Maidstone
Kent, ME17 3WD
Tel: 0845 600 8276
www.textile-recycling.org.uk

Textiles can be recycled in charity shops all over the UK. These include Oxfam, Help The Aged, Cancer Research, TRAID, Scope and many more. Some of these can be found in the NGO directory at the back of this book.

Car Recycling

*Website that will help you find
your nearest Auto Recycler.*
www.car-recycling.co.uk

Tyre Recycling

All the hows whats and whens.
www.tyredisposal.co.uk

Retread Manufacturers Association (RMA)

PO Box 320
Crewe
Cheshire, CW2 6WY
www.retreaders.org.uk

UNITED STATES

Technology Recycles

*Recycles and disposes of
computers, based in Colorado.*
Tel: 303 766 9608
www.techrecycle.com

Computer Recycling Centre (CRC)

3249 Santa Rosa Avenue
Santa Rosa, CA 95407
Tel: 707 570 1600
www.crc.org

Share the Technology

*A free, public service
database on which people
can donate used computers
to non-profit organisations.*
PO Box 548
Rancoas, NJ 08073
www.sharetechnology.org

Electronics Recycling

*A nationwide database for
electronics recycling, listing
all information about who to
contact and how to recycle.*
Contact via e-mail only
www.electronicsrecycling.org

NRC Electronics Recycling

*An electronics recycling
initiative run by the National
Recycling Coalition, which
points you in the right
direction for information,
contacts, policies and
programmes for e-cycling.*
www.nrc-recycle.org/
resources/electronics/index

Recycle 4 Dollars

*Recycle your inkjet toner
cartridges and donate
money to charity.*
Tel: 1800 640 3880
www.recycle4dollars.com

Rechargeable Battery Recycling Corporation

Information and assistance about how to recycle your rechargeable batteries.
1000 Parkwood Circle
Suite 450
Atlanta, GA 30339
Tel: 678 419 9990
www.rbrc.org

Used Recycle.net

A web based exchange of large household goods with listings from all over the country.
www.used.recycle.net

Collective Good Mobile Phone recycling

The mobile devices recycling resource – helps you dispose of your mobile/PDA or pager in a responsible manner.
4508 Bibb Blvd, Suite B–10
Tucker, GA 30084
Tel: 770 856 9021
www.collectivegood.com

Tire Re-tread Information Bureau

900 Weldon Grove
Pacific Grove, CA 93950
Tel: 831 372 1917
www.retread.org

Recycled Textiles Association

Information about the textile recycling industry.
7910 Woodmont Ave.
Suite 1130
Bethesda, MD 20814
Tel: 301 718 0671
www.textilerecycle.org

CANADA

Computers For Schools Program

Industry Canada
155 Queen Street
4th Floor
Ottawa, ON K1A 0H5
Tel: 1888 636 9899
www.cfs-ope.ic.gc.ca

Electronics Product Stewardship Canada

All about e-cycling in Canada.
130 Albert Street, Suite 500
Ottawa, ON K1P 5G4
Tel: 613 238 4822
www.epsc.ca

The Rubber Association of Canada

2000 Argentia Road,
Plaza 4, Suite 250
Mississauga, ON L5N 1W1
Tel: 905 814 1714
www.rubberassociation.ca

Canadian Wood Council

99 Bank Street, Suite 4000
Ottawa, ON K1P 6B9
Tel: 1 800 463 5091
www.cwc.ca

AUSTRALIA

Com.it

An IT recycling initiative.
114-118 Campbell Street
Collingwood
Victoria
Tel: 61 3 941 62604
www.com-it.net.au

Green PC

Nationwide organisation refurbishing and selling old computers.
375 Johnston Street
Abbotsford 3067
Tel: 03 9418 7400
www.greenpc.com.au

NEW ZEALAND

The Ark
Computer recycling.
2 Ross Reid Place
East Tamaki
Tel: 09 272 2676,
www.the-ark.co.nz

PC Recycling Channel
27b Dragon Street
Grenada North
Wellington
Tel: 04 232 4285
www.pc-recycling.co.nz

Molten Media Trust
1st Floor, 148 Lichfield Street
Christchurch
Tel: 03 377 1154
www.molten.org.nz

Case study 11

E-waste exportation and exploitation

China

India

A dangerous new waste stream is rapidly emerging. The world's booming consumption of electronic and electrical goods has created a corresponding explosion in electronic scrap containing toxic chemicals and heavy metals, which are notoriously difficult to dispose of.

Every year, thousands of scrap computers and mobile phones are exported, often illegally from the European Union, United States, Japan and other industrialised countries, to the Far East, India, Africa and China. In Western countries, electronics recycling takes place in purpose built recycling plants under more or less controlled conditions. In many EU states for example, plastics from e-waste containing brominated flame retardants and PVC (vinyl) cabling are not recycled to avoid brominated furans and dioxins being released into the atmosphere. In developing countries, however, there are no such controls. Recycling is done by hand in scrap yards, often by children, and these workers are exposed to a cocktail of toxic chemicals and poisons when they break the products apart.

Greenpeace International conducted an extensive scientific investigation into the hazardous chemicals found in e-waste scrap yards, by analysing the dust from workshops and the wastewater, soil and sediment from rivers in the surrounding area. The report, published in August 2005, shows conclusively that all stages in processing the e-waste enable toxic chemicals, including heavy metals, to be released into the workplace and the surrounding environment. Concentrations of lead in dust samples collected from some workshops in China, for example, were hundreds of times higher than typical levels of household dusts. The wastewater and sediment results were equally disturbing, indicating high levels of several heavy metals and residues of phthalates and PCBs (polychlorinated biphenyls) amongst numerous other toxins.

Another environmental organisation, the Basel Action Network (BAN) together with Greenpeace China, visited Guiyua in the Guangdong province, the main centre of e-waste scrapping in China, to find 100,000 migrant workers including children living in devastating conditions. The report found that:

> There are no health or environmental controls. Plastics and wires are burned in the open; soldered circuit boards are melted and burned; lead contaminated cathode ray tubes are dumped. Observers reported seeing women and girls heating lead solder in woks over open fires to loosen memory chips from circuit boards. Afterwards, the used lead, universally recognised as a neurological toxin, is tipped like kitchen slops on to the ground. The residue that has no value is dumped in fields and irrigation canals with the result that local well water is polluted and fresh water has to be trucked in from elsewhere.[13]

The impacts of unregulated e-waste recycling on the health of recycling workers and the surrounding communities is largely unstudied, and it is impossible to know what their long-term effects will be.

The exportation of e-waste to developing countries is often in violation of the Basel Convention, a landmark global convention put forward by the UN in 1989, signed by countries the world over, prohibiting the shipment of hazardous waste to countries lacking in the technical capacity to manage and dispose of it properly. Inspections of 18 European seaports in 2005 found as much as 47 per cent of waste destined for export, including e-waste, was illegal. In 2003, in the UK alone, at least 23,000 metric tons of undeclared or 'grey' market electronic waste was illegally shipped to developing countries. In the US it is estimated that 50 to 80 per cent of waste collected for recycling is being exported in this way. This practice is legal because the US has not ratified the Basel Convention, although clearly it is immoral.

Mainland China tried to prevent this trade by banning the import of e-waste in 2000. However, Greenpeace has discovered that the laws are not working: e-waste is still arriving in Guiyua and there is a growing e-waste trade problem in India. 25,000 workers are employed at scrap yards in Delhi alone, where 10 to 20,000 tons of e-waste is handled each year. Other such scrap yards have been found in Meerut, Ferozabad, Chennai, Bangalore and Mumbai.

13 Girling, Richard, *Rubbish!*, Eden Project Books, 2005, p. 351

Top: Children in a mountain of e-waste.
Bottom: A computer sports the name of its previous owner – unlikely to be Chinese.
Photographs by Natalie Behring.

How did this trade evolve?

In the 1990s, governments in the European Union, Japan
and some US states set up e-waste 'recycling' systems. But
many countries did not have the capacity to deal with the
sheer quantity of waste that they generated or the hazardous
materials involved. They therefore began exporting the problem
to developing countries where laws to protect workers and the
environment are inadequate or not enforced. It is also cheaper
to 'recycle' waste in developing countries: the cost of glass-to-
glass recycling of computer monitors in the United States is
$0.50 (0.29 pence) per pound compared to $0.05 in China.

Demand in Asia for electronic waste began to grow when scrap
yards found they could extract valuable substances such as copper,
iron, silicon, nickel and gold, during the recycling process. A mobile
phone for example is 19 per cent copper and eight per cent iron.

New laws in Europe and Japan are shifting responsibility for
e-waste away from taxpayers, local authorities and governments
and onto the manufacturers of the products. The laws also
ban the use of certain hazardous substances. In response,
companies with EU and Japanese markets are substituting some
toxic substances with safer alternatives and redesigning their
products to make them easier and safer to dismantle and recycle
when they are discarded and returned to them. As of August
2005, the Waste from Electrical and Electronic Equipment
(WEEE) Directive, passed in November 2002, makes producers
responsible for taking their e-waste back when their products
are discarded and sets targets for reuse and recycling.

Whilst these policies indicate a step in the right direction,
Greenpeace has expressed trepidation that unless they are
supported by initiatives that ensure the safe disposal of the products
once they are taken back and sufficient recycling capacity, they
could lead to an increase of e-waste exports to developing countries.

Top: A boy winces in the acrid smoke of
a melting motherboard in Delhi. Photograph
by Prakash Hatvaine.
Bottom: Workers in an e-waste recycling
facility in China. Photograph by Natalie Behring.

Opposite top left: Circuit boards waiting
to be dismantled.
Opposite top right: Parts are burned on open
fires to extract metals.
Opposite bottom: A migrant child from Henan
province recruited for the dirty job of e-cycling
in China. Photographs by Natalie Behring.

Case study 12
TRAID: Designer charity

UK

Top: An employee sorts through mountains of donated clothing.
Bottom: Modelling the final product.

At a boutique in west London, fashionable shoppers pick through rails of designer clothes and accessories. Some items are vintage, classic designs that never go out of fashion. It looks like many of the other fashionable clothes shops in London; the window display is immaculate, the décor tasteful; dance music plays softly in the background.

The difference is that this is one of the eight TRAID shops dotted around London and Brighton. Its initials standing for 'Textiles for Aid and International Development'; everything sold by TRAID is 100 per cent recycled. Even the garments that appear brand new have been fashioned using second hand rather than virgin materials. The money that TRAID makes from selling these clothes is spent on supporting sustainable development in some of the poorest countries in the world.

Selling second hand clothes for charity is not a new idea. Charity shops and thrift stores have been around in one form or another for a long time doing just that. They provide a valuable service, collecting and reselling unwanted clothing and household items that would otherwise be discarded, thus reducing waste and raising money for those in need. They are an important source of cheap affordable clothing and they also collect donations of garments to send overseas. For the eagle-eyed and fashion conscious they are a good place to snap up bargains and vintage clothing. But charity shops in England can often appear more functional than funky. The founders of TRAID (started in July 1999 at the recommendation of the Charity Commission) saw potential in the idea of re-selling quality recycled clothing in a chain of boutique type shops in London and Brighton. They believe that by "recycling textiles and retailing them with a fashion edge" more value can be extracted from the raw materials.

At TRAID's warehouse in Wembley, London, second hand garments and fabrics are piled up high, ready to be reused. Most of this second hand clothing and material is donated by members of the public, placed into one of TRAID's 750 recycling bins that are located at sites across the UK such as supermarkets and pub car parks. The organisation also works in tandem with local councils, paying for and installing textile recycling banks at locations easily accessible to the local community.

This is a mutually beneficial arrangement; TRAID is able to source garments and fabrics to sell in its shops, and because all textile donations are weighed after they arrive in Wembley, TRAID can give each local authority an accurate figure of how much material has been collected and recycled from their constituency. This helps them to reach their yearly recycling quotas. TRAID also combats the traditionally wasteful practices of high street fashion retailers by accepting donations of clothes and fabrics that are deemed unsuitable for sale or surplus to requirements and might otherwise be destined for landfill.

TRAID recycling bins are emptied once a week. All the donations are collected and driven back to the warehouse in Wembley. After weighing, the sorting begins. Shoes, clothing and fabrics are all placed onto a conveyor belt and are picked through by trained sorters and organised into particular categories. Some items are suitable for immediate resale at one of the eight TRAID shops. Each of these shops are designed to reflect the style of the local community; saris and silks are sent to the Kilburn branch, at Westbourne Grove the emphasis is on designer vintage wares and in Holloway, second hand clothing is sold at especially affordable prices. TRAID employs knowledgeable staff to fine-sort all the garments and ensure that they end up at the right shop.

Some of the donations are not suitable for immediate resale. TRAID send these items to their team of freelance fashion designers who create new items of clothing out of the old garments and lengths of fabric.

Various techniques are used to produce the TRAID Remade range of clothing. The designers must be creative with, and reactive to, the material they receive. Suit jackets are customised, detail is added through prints or stitching and a new item of desirable clothing is created out of an old garment; unfashionable, damaged leather and suede jackets that would be of little value to a modern consumer are cut up and re-sewn as belts, chokers and handbags; old curtains, table cloths and items of clothing are also cut up, to become dresses, shirts and shopping bags. The original material is cut and spliced with other recycled fabrics to produce something completely new. TRAID Remade produces around 20 items of clothing each week. The new garments and accessories are sold alongside the second

Top: One of TRAID's clothing banks.
Middle: A truck picks the clothes from the bank and takes them to the sorting warehouse.
Bottom: A TRAID shop sells the redesigned clothes onwards.

FROM A SELECTION OF HANDPRINTED MENS SWEATSHIRTS AND JACKETS

traid REMADE

FROM A SELECTION OF APPLIQUED AND HANDPRINTED
RE-SHAPED KNITWEAR AND SWEATSHIRTS

hand clothing in TRAID's shops and through some outside stockists. TRAID believes that poor countries are best helped through long-term sustainable development projects. Over the last five years TRAID has donated over £700,000 ($1.25 million) to ten long term projects and £135,000 ($250,000) to emergency appeals. Current projects include funding education in Afghanistan, health and sanitation in Madagascar and the prevention of domestic violence in Costa Rica.

From a recycling point of view TRAID diverts over 2,500 tons of textiles from landfill every year, reusing 94 per cent of the donations they receive. TRAID also runs education projects in schools that teach children how to reuse and be creative with old clothes. In order to reduce plastic bag waste, TRAID charges ten pence for their plastic bags and alters a trendy alternative in the form of a 'Bag for Life.' This reusable shopping bag, designed by Wayne Hemingway of Red or Dead is made from durable natural fibres and is on sale both in their shops and via their website. At the moment TRAID only has shops in London and Brighton but they are hoping to expand across the UK. This will mean that their quirky, innovative designs are more widely available, and at the same time will increase awareness of textile reuse, save more textile waste from landfill and mean that more money can be donated to worthy causes.

Case study 13

UK

From cradle to grave: The nappy debate

People often start thinking about the future of the environment when they start having children. After all, an ozone layer riddled with holes, a planet heating up like a microwave, and an endless expanse of non-biodegradable waste is not a great legacy to leave. For this reason, the issue of reusable versus disposable nappies has evolved into a large-scale debate, reaching its climax in May 2005, when the Environment Agency of Britain published a 210 page report stating that there was no substantial difference between reusable and disposable nappies. Before we look at this debate in more depth, exactly what are the issues at stake?

Nappies are generally used in the first two and a half years of a baby's life. By this age, approximately 90 per cent of girls and 75 per cent of boys have complete bladder control. Until that point in time, they will produce approximately 254 litres of urine and 98 kilograms of faeces. What this equates to on a daily basis is around 4.16 nappies per day. Or, over the whole two and a half years, 3,796 nappies. In weight this is equivalent to 169.5 kilograms of nappies that need to be purchased – and that's before soiling. Including urine and excreta, the weight shoots up to 537.6 kilograms – that's half a million tons of waste for Britain annually. So we're not talking small quantities.[14]

The Report

The Environment Agency in the UK set up an extensive and expensive survey to find out whether the alternative option of reusable nappies was environmentally advantageous. In order to do so, they conducted a detailed life cycle analysis (LCA) of the environmental impacts of the manufacture, use and disposal of reusable and disposable nappies.

14 All statistics from: "Life Cycle Assessment of Disposable and Reusable Nappies in the UK", Environment Agency Report, May 2005

The impacts assessed were as follows:
- Global warming
- Ozone depletion
- Summer smog formation
- Depletion of non-renewable reserves
- Nutrient water pollution
- Acidification
- Human toxicity
- Aquatic and terrestrial toxicity measures

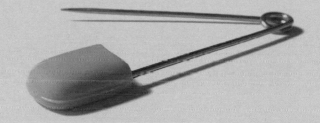

Disposable nappies

There are three main manufacturers and suppliers of disposable nappies in the UK: Proctor and Gamble; Kimberly Clark; and SCA Hygiene. They sell approximately 2.47 billion nappies per year, virtually all of which are landfilled. This equates to three per cent of household waste in Britain; the largest single product category found in household waste. Disposing of these quantities is costly. Kent councils estimate that they spend over £2 million ($3.5 million) a year dealing with nappy waste.

Most disposable nappies sold in the UK are also manufactured in the UK, with a few raw materials imported from abroad. The main components of disposable nappies are cellulose fibres and polymer. Generally speaking, the nappies are transported to retailers, and consumers purchase them from their local retail branch and transport them home for use.

Britain throws away 8 million nappies a day

**Real nappies don't cost the earth
... anything else is just rubbish.**

Women's
Environmental
Network

Women's Environmental Network PO Box 30626, London E1 1TZ.   www.wen.org.uk

March 2003

Reusable Nappies

Reusable nappies account for less than four per cent of the nappy market, although they are gaining in popularity. There are a few different types of reusables, but the general formula is a soft absorbent cotton nappy with a liner, and waterproof, breathable wraps around them. Most of the terry cloth nappies sold in the UK are produced abroad. Consumers purchase the nappies via retail outlets or the internet. During use, terry nappies are often soaked in a solution of sanitising fluid prior to washing. They are then put through a laundry in the washing machine and either tumble or air dried. Reusable nappies tend to be slightly less watertight than disposable ones, and are therefore changed more frequently.

People using reusable nappies either use their home laundry or employ a nappy laundering service – which pick up the nappies and transport them to a centralised laundering service.

The survey attempted to include all significant processes, tracing material and energy flows to the point where material and energy are extracted or emitted to the natural environment. So things like the average emissions used when driving to the supermarket; the energy used to harvest the cotton and transport it to the ginning plant; and the quantities of energy used by a tumble dryer have all been taken into account. The report concluded as follows:

- For disposable nappies, the main impacts relate to manufacturing including raw material production and waste management.

- For home use reusables and commercial laundering, the main source of environmental impact is from generating the electricity used in washing and drying.

There was little or no difference to environmental impact between the different types of nappy; global warming and non-renewable resource depletion impacts over the two and a half nappy-wearing years were comparable with driving a car between 3,000 and 5,600 kilometres.

The final section of the report concluded that disposable nappy manufacturers should focus on weight reduction and improvements in manufacturing materials, whilst reusable users should focus on using less energy in washing and drying.

Women's Environmental Network

The Debate

Environmental organisations responded ferociously to these findings. As Nicola Baird of Friends of the Earth stated, the report only looks at the chemical effect on the environment of the different nappies, it does not look at the huge volume of disposable nappies that end up in landfill; "If James I had been wearing nappies – over 400 years ago – they'd only just be rotting down now. That's a lot of landfill."[15] The Women's Environmental Network (WEN) went further and stated that the statistics on which the report was based, were flawed. Ann Link of WEN states: "If parents use 24 nappies and follow manufacturers' instructions to wash at 60 degrees Celsius using an A rated washing machine, they will have approximately 24 per cent less impact on global warming than the report says."[16]

WEN also notes that whilst 2,000 parents using disposable nappies participated in the survey, only 117 parents using reusable nappies were included, and as there are a number of different types of reusable nappies, the result was that a very limited number of respondents were relied upon for some of the key assumptions.

In the concluding section of the report, the Environment Agency recommends that disposable nappy manufacturers reduce the weight of their nappies; but this contradicts the section in the report noting that the super-absorbent polymers (which are key in weight reduction) are the largest source of environmental impact for disposable nappies.[17]

Whilst the report is very obviously flawed, its underlying point still holds true; companies and parents could and should be doing more to protect the environment. Parents who insist on using disposable nappies should look at buying 'eco-disposable nappies', which use recycled product and unbleached pulp (available at around 12 pence per nappy from www.spiritofnature.co.uk), and parents using reusable nappies should attempt to cut down on energy consumption of the washing and drying process.

15 Quoted in Flanagan, Ben, "Turn Your Baby Green", *The Observer,* 22 May 2005

16 WEN Media Statement "Environment Agency nappy report is seriously flawed", 19 May 2005, available on www.wen.org.uk

17 WEN's comments on LCA section 8.1.2, Disposable Nappy Manufacturing

WEN has published a list of hints and tips for reusable nappy users to take into account:

Use an A rated washing machine — achieving a 14 per cent reduction in global warming impact.

Wash wet nappies at lower temperatures along with the rest of your laundry.

Don't tumble dry — air dry.

Don't soak — store nappies dry in a lidded bucket.

If soaking, don't use a sanitiser — use a natural agent.

Don't use conditioner — it reduces absorbency and is an unnecessary use of chemicals.

Use an eco-detergent.

Use washable liners.

Never iron nappies or wraps.

Avoid PVC wraps.

Use organic products.

Use second hand nappies. Extend the life of your nappies — reuse them on another baby, or give/sell them to someone else.

We shall never understand the natural environment until we see it as a living organism. Land can be healthy or sick, fertile or barren, rich or poor, lovingly nurtured or bled white. Our present attitudes and laws governing the ownership and use of land represent an abuse of the concept of private property.... Today you can murder land for private profit. You can leave the corpse for all to see and nobody calls the cops.

Paul Brooks, The Pursuit of Wilderness, 1971

Compost

Over a third of your bin is made up of organic content. All the food scraps, potato peels, coffee grounds, eggshells and tea bags that you casually dispose of, end up in landfills taking up vast amounts of valuable landfill space and slowly decomposing, emitting methane gas into our already overheating atmosphere and leaching acid into the soil.

Compost is nature's way of recycling. It is the transformation of organic waste through decomposition into a rich soil-like material. When it is added to gardens, compost conditions the soil, improving its structure, texture and aeration. It contributes to erosion control, soil fertility and healthy root development in plants by returning nutrients such as phosphorus, potassium, magnesium, zinc and iron to the soil.

The process of composting is the process of decomposition. When a plant dies, its remains are attacked by micro-organisms and invertebrates in the soil until it decays. The decomposition process can be speeded up by optimising the living conditions for the micro organisms. This means providing the right amount of food, water and oxygen so the micro-organisms can do their job. In practice this means turning the compost pile regularly to provide aeration, keeping an eye on the level of moisture of the compost heap and maintaining the correct carbon to nitrogen ratio. This last factor might sound a bit scary but what it actually means is that certain organic materials, such as paper, leaves and woodchips, are high in carbon, whilst grass clippings and manure are high in nitrogen. As microbes like their food high in carbon and low in nitrogen, in practical terms it means, don't overload your compost heap with grass clippings.

Centralised Composting

Centralised composting involves the collection of yard waste via a kerbside scheme and transportation of the waste to a special facility, which processes it into compost. These facilities are designed to manage large volumes of organic materials. They form the organic waste into windrows, long triangular shaped rows reaching up to two and a half metres high. Once a month, the rows will be turned for aeration, using a front-end loader. The temperature and moisture are checked twice a week. The finished compost is either sold or given away. Centralised systems are a good solution for restaurant, supermarket and grocery store waste. The main challenge facing this type of composting is ensuring the organic purity of the product. If glass gets mixed in with organic waste, it is very difficult to screen it out, and the resulting compost is difficult to sell on.

Backyard Composting

If there is no centralised composting centre in your vicinity, you can easily make your own compost in your back garden or even in a flat. This is in fact more efficient than centralised composting as no heavy machinery is required.

Composting your kitchen and garden waste is easy

Put these in

Keep these out

How to compost:

Compost systems can be made very easily, by building a rudimentary enclosure of wood or old tyres, covered with a sheet of plastic or an old carpet. Alternatively you may wish to buy a commercial unit, often available through your municipality or to order on-line. Bear in mind that compost decomposes faster if it covers a larger surface area.

The most important thing is to place your composter in a sunny area with good drainage.

- Turn the soil where the composter will be.

- Cover the base of the composter with small branches to allow for air circulation and drainage.

- Alternate wet waste (table scraps) and dry waste (garden refuse).

- Add some finished compost or garden soil to the mixture to get it started.

Above: Various types of homemade compost enclosures.
Right: A store-bought composting unit.

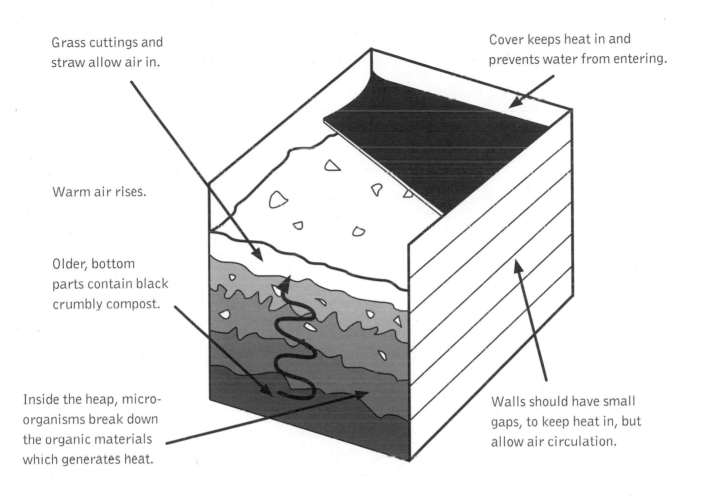

Grass cuttings and straw allow air in.

Cover keeps heat in and prevents water from entering.

Warm air rises.

Older, bottom parts contain black crumbly compost.

Inside the heap, micro-organisms break down the organic materials which generates heat.

Walls should have small gaps, to keep heat in, but allow air circulation.

⟳ Do compost

Grass clippings, leaves, twigs, pine needles, sawdust, straw, shredded newspaper, cardboard, paper towels, vegetable and fruit scraps, bread, eggshells, teabags, coffee grounds, hair and lint and farm manure.

Do not compost

Big branches, painted wood, sawdust from plywood, coloured or coated paper, meat, fish or dairy products, diseased plants or pet waste and litter.

Make your own wormery

If you do not have a garden, vermicomposting, or worm composting, may be the answer. Using tiger or brandling worms (available from fishing shops), they only produce a small quantity of compost and a liquid which forms a concentrated plant food. You can buy a worm bin online or from your municipality, or you can make one yourself:

2 Place seven centimetres of sand or gravel at the bottom.

1 Get a plastic dustbin and drill some holes in the lid.

3 Place some thin wooden slats on top of the gravel.

4 Place some bedding on top of the slats – this can be shredded cardboard and/or paper that is moistened to about 75 per cent water content.

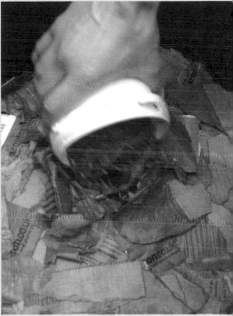

5 Drill a plastic tap into the bin, just at the level of the wooden slats. You can buy a tap from most hardware stores.

6 Dig a small hollow in the bedding material and place 400 or so red worms inside.

7 Then add your food scraps, making sure they are cut up reasonably small (see the How to Compost section for 'do's and don'ts' of table scraps). Cover the surface of the bedding with food, but leave a small gap so the worms can get out if they need to.

8 Cover the result in a thick sheet of wet newspapers.

Only add more food when the worms have finished their last lot! In a healthy box, worms can consume two to three kilograms of food scraps per week.

About four to six months after the box has been started, the worms will have converted all of the bedding and most of the food into 'castings', which will need to be harvested so the process can start again. These castings make great plant food. Sprinkle them on your plants to help them grow.

Composting with EM Bokashi

The Japanese word 'Bokashi' means 'fermented organic matter' and has been used since ancient times as fertiliser. More recently, the recipe has been developed by Japanese university professor, Teruo Higa, to include EM. EM means Effective Micro-organisms. It consists of mixed cultures of naturally occurring, beneficial micro-organisms possessing antioxidant qualities, such as lactic acid bacteria, yeast, photosynthetic bacteria and actinomycetes. It's a bit like the good stuff you might find in a probiotic drink — but very highly concentrated. To make EM Bokashi, the micro-organism cultures are mixed with a combination of molasses, water and wheatbran or coffee husks, and left to ferment in an airtight container for six months. The resulting mixture has a dry, flaky texture, a bit like fish food.

Composting with EM Bokashi is probably the easiest and most effective way of making compost, especially if you are in an urban environment. The micro-organisms aid the fermentation of the organic matter and prevent rotting — it's a kind of pickling process. In this way the composting process can be accelerated by 50 per cent, and bad odours can be totally eliminated, along with fruit flies and other nasties that unaided composting sometimes incurs.

You can order Bokashi from some of the websites listed in the back of this book. Use it as follows:

- Put a little Bokashi at the bottom of your kitchen compost bin.

- Every time you add kitchen scraps, sprinkle a layer of Bokashi on top (one handful for every three centimetres of food).

- Using a plastic bag, compress the waste to remove air, every time you add more scraps.

- Replace the lid tightly after every use.

- If there seems to be an excess of liquid, drain this off and use as a fertiliser.

- When your bin is full, add a generous layer of Bokashi.

If your bin smells lightly of vinegar, that's a good indication that the fermentation process is occurring.

Once your bin is full, bury the contents in a hole about 20 centimetres deep. It's acidic when you first put it in, so keep it away from direct contact with plant roots. Bacteria in the soil and compost will start to break down the food and it will neutralise after seven to ten days. After two to three weeks, the food will have decomposed in the soil and will serve as a rich nutrient like any other compost.

Mulching

Mulching is another form of composting, and basically means a layer of organic material on top of the soil. It is a slow but efficient form of lawn maintenance and weed control.

The easiest way to mulch is simply not to bag your lawn clippings. This will reduce water evaporation from your lawn and facilitate better growth by keeping the soil temperature cooler. Eventually the clippings will work their way back into the soil, enriching it in the same way that compost does.

A few tips

- Don't let your grass grow too long before mowing. The clippings should be no more than two centimetres long.

- Don't overfertilise your lawn. Too dense growth will not allow the clippings to reach the soil.

- Mow your lawn when it is dry so the clippings can filter down to the soil more easily.

- You may choose to collect your clippings every third time you mow, so as not to overload the lawn and prevent sunlight from reaching the grass. Use these clippings for your compost heap!

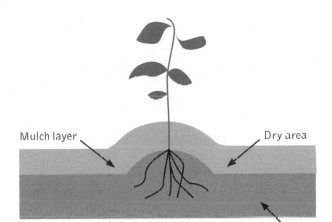

Mulch layer   Dry area

This area is moist. Moisture moves up to the dry area.

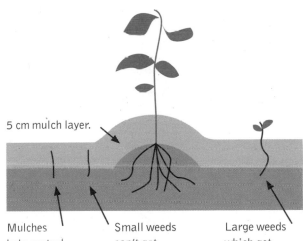

5 cm mulch layer.

Mulches help control nutgrasses and Johnson grass.

Small weeds can't get through the mulch layer.

Large weeds which get through are easily pulled.

The final product.

UK

Community Composting Network (CCN)
Information about setting up a composting scheme in your community.
67 Alexandra Road
Sheffield, S2 3EE
Tel: 0114 258 0483
www.communitycompost.org

The Composting Association
Avon House
Tithe Barn Road
Wellingborough
Northamptonshire, NN8 1DH
Tel: 0870 160 3270
www.compost.org.uk

HDRA
The organic organisation gives instructions and tips on composting.
Ryton Organic Gardens
Coventry
Warwickshire, CV8 3LG
Tel: 0247 630 3517
www.hdra.org.uk

Straight Recycling Systems
Sells compost bins and other products.
31 Eastgate
Leeds, LS2 7LY
Tel: 0113 245 2244
www.straight.co.uk

Primal Seeds
More information on how to compost.
Contact via website and e-mail
www.primalseeds.org

Living soil
A website pioneering sustainable communities, from which you can buy Bokashi.
www.livingsoil.co.uk

UNITED STATES

The US Composting Council
4250 Veterans Memorial
Highway, Suite 275
Holbrook, NY 11741
Tel: 631 737 4939
www.compostingcouncil.org

EPA Composting page
See directory for contact details.
www.epa.gov/epaoswer/
non-hw/composting

NYC Compost Project
Recycling in New York City – how to, links and resources. Funded by NYC Department of Sanitation.
www.nyccompost.org

Sustainable Community Development (SCD)
A Missouri based website dealing in sustainable agriculture. Supplies Bokashi.
www.scdworld.com

Worm Woman
All your worm composting needs addressed.
10332 Shaver Road
Kalamazoo, MI 49024
Tel: 269 327 0108
www.wormwoman.com

CANADA

Composting Council of Canada
Great resource – everything you need to know, well laid out and very passionate!
16 Northumberland Street
Toronto, ON M6H 1P7
Tel: 1 877 571 GROW
or 416 535 0240
www.compost.org

NEW ZEALAND

University of Canterbury
They run a composting operation and highly interactive website. Ask the professor all your composting questions.
Contact via website
www.civil.canterbury.ac.nz/compost

Case study 14

New York state prisons' composting coup

USA

New York City has a serious rubbish problem. Fresh Kills, the landfill opened in 1948 to service the metropolis, was closed in March 2001 under pressure from the Environmental Protection Agency and the residents of Staten Island. The landfill is one of the largest refuse heaps in human history, and is now due to be transformed into wetlands, recreational facilities and parkland. This, however, does not solve the Big Apple's big problem, and New Yorkers are currently paying $70.00 (£40) per ton to have their rubbish shipped out of state. Not that the mayor seems to feel any pressure – or in fact the local population. Recycling rates reach a mere 18 per cent and for two years between 2002 and 2004 there were no glass or plastic recycling facilities at all – as they were deemed too expensive.

Oddly, the New York State Department of Correctional Services (DOCS) seems to be the one place where recycling is given due respect. In 1989 a survey found that food scraps comprised 30 per cent of the waste stream in prisons. In an attempt to reduce disposal costs, DOCS began a large-scale composting programme; with 32 composting facilities servicing 54 of the 70 correctional facilities in the state. Every day kitchen workers and inmates dispose of excess food (including meat) in unlined plastic containers. These are refrigerated to prevent putrefaction, until inmates take them to large composting sites, three or four times a week. These sites are set up in windrows, which the inmates are also responsible for looking after. They bulk up the compost with wood chips, produced inside the prisons, and some programmes also use yard debris from neighbouring communities. A steady temperature of 64 degrees Celsius is maintained by means of a strict regime of bulking ratios and turning frequencies.

Out of a total population of 65,000 inmates, about 1,200 are involved in these operations. They receive full training, which often helps them find jobs in recycling and composting industries upon their release, and three of the facilities offer technical training that provides a more in-depth, scientific approach. Most of the inmates have responded positively to this process, and feel that it is a productive endeavour, which yields tangible results. The compost that is produced is used for the landscaping of the prison grounds, and the 20 per cent left over is supplied free to local communities.

However, the resource management director, Jim Marion, is quick to point out that the reason the programme was implemented was not for the environmental benefits, nor for the good of the prisoners, but in order to cut costs. An aim it has achieved admirably. What started out as a small, low-tech, state funded experiment, has proved itself above and beyond all expectations. Despite increased hauling and tipping rates, DOCS garbage disposal costs decreased by 10.3 per cent in the first seven years of its composting programme. It works out that the average composting costs are $34 (£20) per ton, whereas landfill hauling and tipping fees work out at $125 (£72) per ton, which makes a saving of $91 (£52) per ton. Jim Marion says: "When one 1,500 inmate facility started their composting operation, their first month the garbage bill went down by $3,500 (£2000). Last year our avoided cost was more than we paid for waste disposal within the entire department."

Similar prison composting programmes are being implemented in Virginia, using composters rather than windrows, and the benefits are applicable to prison services worldwide.

Top: Open windrows.
Middle: Aerated bays.
Bottom: Open windrows with Easter
Correctional Facility, Napanoch,
New York in the background.

Case study 15

![Japan flag]

Japan

The Regional Circulation Network — food waste recycling in Nagaoka City, Japan

Unlike the extended producer responsibility initiative in Germany, Japan regulates its waste stream by means of shared responsibility between government, industry and consumer. Funding and support is given to companies and organisations that develop technologies for waste treatment and recycling, and a waste management convention, WASTEC, is held every year to coordinate efforts.

The winner of the Environment Minister's Award prize (the grand jury prize of WASTEC, as it were) at the 2004 convention, was a small NGO called the Regional Circulation Network, based in Nagaoka, a small city to the north of Tokyo. The NGO started out as a citizen's group called the 'Mizubasho', which was formed in 1994 to focus on recycling household foodwaste, and moved on in 1997 to recycling food in schools. It currently covers 88 schools and recycles approximately 1,000 tons of waste every year.

The organisation works by giving scraps from school lunches to livestock operations in the area. Employing only nine permanent members of staff, it relies on the volunteers and the local community to assist in its efforts. Cooks and students sort and drain the food waste and divide it into two categories – vegetarian (for cattle) and meat (hogs and minks). The waste is sprinkled with EM Bokashi (see p. 186), and mixed with cracker crumbs to absorb moisture. In this way, the mixture is dried, and reduced in weight. Another week and the fermentation process is complete, and the resulting concoction can be used as feed for livestock.

The Regional Circulation Network is now expanding
its initiatives to include such facilities as chopsticks
recycling (to make paper), cooking oil recycling and an
exchange programme where people can swap properly
dried and treated kitchen scraps for meat and eggs.

リサイクル介護用品支援センター

Case study 16

UK

Hackney's Nightingale Estate: Composting but not as we know it

At the Nightingale Estate in Hackney, The East London Community Recycling Project (ELCRP) composts household food waste from ten estates all over Tower Hamlets. Given the size of the project, one might anticipate giant haystack-sized piles of rotting vegetables or, at the very least, industrial vats of food in varying states of decomposition. However, the composting facility at the estate is very different to what might be expected. The air is fresh in the August sun, and in a covered area there are three quietly whirring, sleek silver machines slightly bigger than office desks. They are labelled Rocket 1, Rocket 2 and Rocket 3 and they precipitate the speedy transformation of food waste collected from the estates into compost.

Futuristic composting machines aside for a moment, the ELCRP are also remarkable because of the obstacles that they have overcome in order to set up the project. There was a real problem with the way waste was being disposed of on the Nightingale Estate, as Cam Matheson, one of the six people who set up the ELCRP in 2001, explains: "In the summer it stank, it really smelt of decomposing food. Waste was piled up beside outdoor rubbish bins. Rats were a common problem." Finding an environmentally sound solution to this problem was complicated by the fact that up until recently collecting household waste for composting was illegal under UK law: "Three years ago what we are doing, and what anyone was doing when they composted meat and fish was illegal. Because of BSC and the foot and mouth crisis, composting laws were changed. The government put in place legislation that assumed that composting would and did spread foot and mouth."

The Animal By-Products Order (as the legislation was called) stipulated that if vegetable matter came into any contact with any meat product then it must be classified as animal

Cam Mattheson at the thrift store
and computer refurbishment
office run by ELCRP.

waste and therefore could not be composted. This meant that any household food waste collected by a local authority would have to be incinerated, there being no way of proving that it was not contaminated in this way. To Cam this would represent "a hell of a waste of a terrific raw material. To burn it or divert it to landfill would terrible."

It was the discovery of 'the Rocket' that allowed the ELCRP to challenge this ruling. The Rocket is a large, mechanised composter that uses what is known as a 'continuous process'. Using a minimal amount of energy, it heats the compost to a temperature of 60 degrees Celsius, constantly turning over the compost, to maintain aeration. By creating these ideal conditions, the bacteria and micro-organisms naturally present in all organic matter are encouraged to break down the food at a fast pace – and safe compost can be formed in a mere two weeks. The Animal By-Products Order did not recognise continuous process as a method of composting but the ELCRP lobbied them to accept that the process was safe, and that by maintaining a consistently high temperature, dangerous bacteria would be eliminated. In Cam's words: "This tiny little organisation at the Nightingale Estate in Hackney actually changed a major piece of legislation."

Having found a method of safely processing household food waste, the other major challenge facing the ELCRP was convincing residents on the housing estates in Hackney that composting is worthwhile. "What ELCRP did (and we were the first in the country to do so) is that we managed to get a system in place that looks after residents. If someone lives on the tenth floor and she is not going to be prepared to take part, if the scheme is inconvenient, then the project will not work. You just antagonise the residents."

The solution to this problem lies in organisation. Estates have a very different streetscape to other residential areas; a streetscape that was not taken into account when national government set local authorities 'Recycling and Recovery' targets. The borough of Hackney is made up of 56 per cent housing estates. These estates often comprise tower blocks several stories high, and enclosed residential areas not easily accessible to vehicles. A kerbside collection scheme that would work in a suburban area is simply not an option. The ELCRP began collecting household waste on a door-to-door basis in 2001, providing residents with green bags in

Top: Wormeries.
Middle: The Rockets, quietly churning the compost.
Bottom: Interior view of the compost inside a Rocket.

which to place all their recyclable waste and then sorting through it themselves. This system is convenient for the residents, as they do not have to carry their recycling bags down flights of stairs and then sort the waste into separate recycling bins. It also allows the ELCRP to collect waste efficiently; the bins are monitored and only taken away when they are full. The money saved on transport and unnecessary collections can be re-invested back into the manpower needed to provide the door-to-door collection service, and the current weekly costs amount to less than 60 pence ($1) per door.

With this network already in place, it was only a small step to provide each resident with a lockable composting bin in which to place their household food waste for weekly collection. However, Cam says that residents might still have been reluctant to take part in the scheme because of the common belief that compost bins are "smelly, they are slimy and that they attract flies, maggots and rats".

This is not necessarily the case, by any means. The ELCRP uses EM Bokashi, the microbiotic system used to prevent putrification (see p. 186). Liquid Bokashi is imported from Holland and mixed by the ELCRP with bran and molasses to make it into dry flakes. The end product looks rather like sawdust. Each resident is given a bag of Bokashi with their composting bin, and are told to scatter a handful of it over their food scraps after they have been placed in the bin. It is a very simple, convenient process, and apparently most residents do follow this system. After all, as Jane the ELCRP's composting manager explains, "it will really stink if they don't! The food will decompose."

The bins are emptied weekly and the food waste is placed straight into the Rocket. After the waste has been composted by the Rocket, it is given back to the residents for use in their gardens and window boxes or used to supplement the wormeries, which run alongside the Rockets producing very high quality fertiliser. Participation rates on the Nightingale Estate have reached around 84 per cent and conditions have certainly been improved. These days it seems tidy and well kept. The ELCRP collects waste from other estates in Hackney and is being closely monitored by other local governments interested in implementing similar schemes in their constituencies. ELCRP are currently looking into the possibility of providing local schools with their own Rockets and Cam suggests that it could also be possible for restaurants to compost their own organic waste in this way.

The ELCRP has also been involved in other initiatives on the estate. Neglected gardens belonging to senior citizens have been rejuvenated, the ELCRP has built them raised beds out of recycled timber and provided fertiliser from the wormery; there is a 'Recycling Shop' where residents can donate or exchange their unwanted household goods; a 'Bicycle Workshop' repairs spare or damaged bicycles and there is a workshop that teaches residents how to use computers run by volunteers using donated computer equipment. The ELCRP furthers its confidence within the community by employing most of its 34 members of staff from the local unemployment register, providing both jobs and training. All of these projects are run with the express intention of involving the entire community in schemes that will improve living conditions for everybody on the estate, and if that isn't enough the Nightingale also has its own 'Recycled Radio' programme broadcast from the estate by Cam and available across London. He spins old jazz and soul records and covers issues close to the heart of the ELCRP.

Left: Jane checks one of the wormeries.
Opposite, clockwise from top: Finished compost. The interior of the wormery. The computers ready to be refurbished.

COMPOST
READY FOR USE
COMPOSTED IN WEBBS ROCKET
PROVIDES IMPORTANT
ORGANIC MATTER AND
FERTILITY TO SOIL

The practice of conservation must spring from a conviction of what is ethically and aesthetically right, as well as what is economically expedient. A thing is right only when it tends to preserve the integrity, stability, and beauty of the community, and the community includes the soil, waters, fauna, and flora, as well as people.

Aldo Leopold

Sustainable design directory

One of the main imperatives for designers and consumers today is to consider the impacts of the products they make and the processes they use on the environment. In order for recycling to compete with virgin products, there needs to be a market for it. As has already been mentioned, it's crucial that we as consumers start ensuring that we print on recycled paper, buy binbags from recycled plastics and use more eco-friendly cleaning products with fewer toxic chemicals.

That's all well and good, but what about the other things we buy; the furniture and the handbags, the clothes and the countless accessories that we've come to rely on? Green design is traditionally associated with overtly hippy aesthetics – hemp shirts and comfortable shoes immediately spring to mind. Not that there's any problem with this, but more modern consumers are often looking for something a bit more, well, trendy, and designers are rising to the challenge. More and more often, designers and consumers are asking themselves some pertinent questions about the products they create or buy: Are the materials used recycled? Are they recycleable? Are the manufacturing processes toxic? Is the final product durable? Is it energy efficient?

Does it require a minimum of packaging? Can it be upgraded as technology advances? The more inventive responses to these questions can be found in the following pages, where we list a range of modern designs from around the world that we thought were particularly innovative or inspiring.

Some of these products are not entirely sustainable – aspects of their production process are not exactly ideal or material types could be better selected. However, we feel that all the products listed provide alternatives that are much more preferable to their mass-produced counterparts By buying eco-conscious design, we as consumers are supporting an evolving mentality, as well as making a statement against unethical production.

From coin purses made out of ring pulls to countertop surfaces made of mobile phones, the following pages will give you a sense of the products that are out there and how to get hold of them.

GENERAL ECO-DIRECTORIES and some shopping pointers.

UNITED STATES

www.ecodirectory.com
A well designed website with sections on products, recipes, book reviews and art.

www.treehugger.com
This is the ultimate in eco-websites. It's clearly written, well laid out and will tell you all you need to know about the products that are out there. It's primarily US-centric, but has entries on innovations in eco-living and recycling from the world over.

www.happyhippie.com
A thorough recycling directory. Reviews are less detailed than those on Treehugger, but still a good resource.

www.greenpeople.org
This claims to be the world's largest directory of eco-friendly products. It is certainly large, and boasts over 7,000 listings.

www.ecomall.com
A good-sized directory with clearly laid out categories. Useful if you can ignore the naff design.

www.twokh.com
An online and physical shop in New York called II KH that focuses on luxury eco-friendly products

www.vivaterra.com
Another online shop that specialises in high-end eco-design. Products have a classic feel.

UNITED KINGDOM

www.ecostreet.co.uk
A web directory for sustainable living. Not enormous but a good few hundred sites are listed – all UK based.

www.naturalmatters.net
"If you want to live in harmony with, and respect for the world, Natural Matters connects you with people, products and ideas for your health, home, pets and more."

www.ecozone.co.uk
A small webshop with a limited but good range including detergent-free eco-balls for laundry and a wind-up mobile phone charger.

www.insightecostore.com
A good webshop with a broader range – and some really fun and funky furniture.

www.recycledproducts.org.uk
This is the website run by WRAP, which provides comprehensive links to recycled products in all sectors – commercial, industrial and domestic. It is constantly being updated, so definitely worth a look.

AUSTRALIA

www.ecoshop.com.au
An online directory, shop and news room for all things eco-friendly.

Bags

Freitag started the trend, but now everyone's doing it. Designers the world over are turning tyres, billboards, inner tubes, seatbelts and advertising banners into colourful, funky bags that the most fashion-conscious of us would be happy to be seen with. Here are some of the best:

Freitag

A Swiss company that started making bags made of used truck tarpaulins, seatbelts and inner tubes in 1993. They are bold with distinct blocks of colour and a slightly skater-y aesthetic.
Freitag Lab AG
Postfach
8031 Zurich
Switzerland
Tel: 41 43 210 33 33
www.freitag.ch

Harvey's Seatbelt bags

Bags made of woven seatbelts in every colour imaginable.
2907 Oak Street
Santa Ana, CA 92707
USA
Tel: 1 714 435 1585
www.seatbeltbags.com

Vy & Elle

Vinyl billboards are popular banners for advertisers – but not so eco-friendly. These two designers turn them into exciting, vibrant bags – each one totally unique.
867 Valley Road
Menasha, WI 54952
USA
Tel: 888 285 4367
www.vyandelle.com

Vaho Is Trashion

Barcelona based company that also makes bags out of billboard banners, as well as other recycled goods. A very alternative look.
www.vaho.ws

Clothes

Worn Again

Funky trainers made of recycled prison blankets, quilts and men's suits.

Anti Apathy
The Hub
5 Torrens Street
London, EC1
UK
Tel: 0208 418 930
www.antiapathy.org

Wired

Recycled jewellery made of metal and rubber washers, cords, alternator wire and steel cable.
store.virginthreads.com/
wiredrelapse.

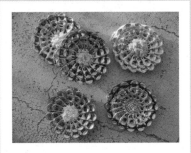

DaLata

Its name meaning 'from the can' in Portuguese, this Brazilian company makes clothes from the ringpulls of cans. Most of the clothes are pretty out there – so for the less risky, there are purses and hats.
www.dalatadesigns.com

Sansegal

T-shirts, fleeces, jumpers and legwear all made from… T-shirts. They have patented a blending system that matches the T-shirts into colour groups before shredding and re-forming them.

611 West 9560 South
Sandy, UT 84070
USA
www.sansegal.com

Furniture

Outdoor Furniture
Durable, hardwearing recycled plastic furniture including benches, walkways, equine and garden products all made of 100% recycled product.
Eco Plastic
86 Annacloy Road
Downpatrick
Down, BT30 9AJ
UK
Tel: 0284 483 1831
www.ecoplastic.net

Thrashcan
A bin made of reclaimed tyres – "Jump, stomp or back over our Thrashcan and watch it bounce back."
www.normalstuff.net/thrashcan

Big Shrimpy Dogbeds
Big Shrimpy makes eco-friendly pet products including fantastic washable, odour resistant recycled fleece dogbeds.
Big Shrimpy
9227 Mary Avenue NW
Seattle, WA 98117
USA
Tel: 206 297 7918
www.bigshrimpy.com

Kwytza Kraft table lamps

Made from post-use disposable chopsticks from restaurants in China – don't worry, they've been thoroughly cleaned first!

372 Florin Road, Suite 307
Sacramento, CA 95831
USA
Tel: 916 760 4188
www.kwytzakraft.com

Segregated recycling and composting bins

Bins that are split down the middle for easy separation of rubbish. Brabantia does a good one, but there are many others, as well as can crushers on this helpful British website.

Home Recycling Ltd
Barwick Lane
Ingleby Barwick
Stockton on Tees
Tees Valley, TS17 5AB
UK
www.homerecycling.co.uk

Futureproof Furniture

A lovely website that specialises in sustainable furniture and accessories.

Hungaria 35 , 1st floor
Vaartkom 35
B–3000 Leuven
Belgium
Tel: 32 16 50 37 50
www.futureproofed.com

Vivavi

A great eco-design website with excellent furniture and slightly less excellent women's clothes.

Vivavi, Inc
644 Manhattan Ave 2nd Floor
Brooklyn, NY 11222
USA
Tel: 866 848 2840
www.vivavi.com

Sculptural Wall Tiles

3-D geometric tiles made from form-moulded post and pre-consumer waste that work with a modern or retro aesthetic. Available from fantastic online furniture design store, re: modern.

932 Pershing Avenue
San Jose, CA 95126
USA
Tel: 408 757 5255
www.re-modern.com

Chatarra

Mercedes Bernardes is an artist and a metallurgist, who creates pieces of furniture and works of art from scrap aluminium (Chatarra means scrap metal in Spanish).

Chatarra
Libertador 15425
B1642CKB – Acassuso
Buenos Aires
Argentina
www.chatarraonline.com.ar

Gueto Design

A Brazilian eco-friendly design company using all types of recycled material, but primarily leather and rubber. All kinds of furniture are available including carpets, chairs, sofas, lamps and benches. Brilliant.

Rod. BR 116, Km
222 no 9250
Caixa Postal 43
CEP: 93950 000
Dois Irimaos
Brazil
Tel: 55 51 564 3131
www.gueto.com.br

David Meddings Design

Eco-friendly indoor and outdoor furniture made of recycled wood as well as home accessories such as soap dishes and coat hooks.

37 St Stephen's Square
Norwich
Norfolk, NR1 3SS
UK
Tel: 01603 629 396
www.reelfurniture.co.uk

Stationery

Recycled Pencils

Green Earth Office Supply sells pencils made using no wood – instead they use recycled blue denim jeans that have been ground up. They also sell pencils made out of recycled dollar bills. Check out their recycled bicycle table and banana fibre notebooks.

Green Earth Office Supply
PO Box 719
Redwood Estates
CA 95044
USA
Tel: 1 800 327 8449
shop.store.yahoo.com/
greenearthofficesupply

Remarkable Recycled

A range of stationery including pencils, mousemats, pencil cases and notepads made of recycled materials including polystyrene packaging, printer parts and recycled plastic cups.

The Green Shop
Cheltenham Road
Bisley
Stroud
Gloucestershire, GL6 7BX
UK
www.greenshop.co.uk

Recycled Paper

A range of recycled paper products including notebooks, paper and craft supplies.

Gate Farm
Fen End
Kenilworth
Warwickshire, CV8 1NW
UK
www.recycled-paper.co.uk

Circuit-board Organisers

A binder made of recycled computer circuit boards filled with recycled paper. Made in the Netherlands and sold in Australia via useful eco-friendly homewares website, Biome.

Biome Living
PO Box 2031
Milton, QLD 4064
Australia
Tel: 1300 301 767
www.biome.com.au

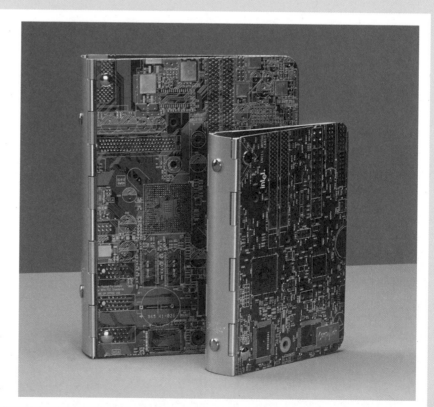

Household goods and miscellaneous

Doormat Flip Flop
A durable, funky doormat made of scrap left over from flip flop production. Available on US based website:
www.wetsand.com

Fairpack
A not-for-profit company that supplies biodegradable plastic bags, recycled paper bags and cotton bags to small retailers at an affordable price. UK based.
www.fairpack.org

Green Glass UK
Lovely, reasonably priced glassware made of recycled bottles. Winner of National Recycling Awards 2001.
Trenant Industrial Estate
Wadebridge
Cornwall, PL27 6HB
UK
Tel: 01208 812 531
www.tradinggreen.co.uk

Animal Bedding
CJ Pet Products do bedding made of 100% recycled material:
PO Box 520
Sheffield, S10 4AW
UK
Tel: 0114 232 3535
www.cjpet.co.uk

Crispina blankets
Colourful, modern blankets made of recycled wool.
22 Walker Street
Lenox, MA 01240
USA
Tel: 413 637 0075
www.crispina.com

Compost Crock
An attractive white ceramic compost crock for your kitchen – easy to clean and dishwasher safe. From:
www.ecozone.co.uk

Used 2 Bee
Home products, furniture and accessories made of recycled goods.
66 Burridge Road
Torquay
Devon, TQ2 6LY
UK
Tel: 01803 607 009
www.used2bee.co.uk

The Acorn Urn
A biodegradable urn made of compressed newspaper.

The Arka Ecopod
An aesthetically pleasing biodegradable coffin. Made of the same material as the Acorn Urn with a fitted, feather lined calico mattress. Both available from the UK based Go Green online shop:
www.gogreen.cellande.co.uk

Organic re-useable nappies
Available in various shapes and sizes. All machine washable at 90 degrees. Available as above on:
www.gogreen.cellande.co.uk

Recycline
Toothbrushes, razors, plates and cutlery made of recycled plastic. The toothbrushes are especially good as they come with a pre-paid envelope so that you can send it back and have it re-bristled once it reaches the end of its lifespan.
Recycline Inc
681 Main Street
Waltham, MA 02451
USA
Tel: 888 354 7296
www.recycline.com

Mint tumbler with dots
These are wine, soda and water bottles that have been cleaned, sandblasted and fashioned into unique glassware. US based.
www.uncommongoods.com

Record Bowl
New York designer, Jeff Davis, makes fruitbowls out of 12" vinyl records. You can even choose which genre of record you want for your bowl!
www.modernartisans.com

The Cardboard Lamp and the Glow Brick

Two clever forms of lighting design – the first made from recycled cardboard compressed into a laminate, and the second is a light that charges up from sunlight or from a lightbulb. Both are available at:
Insight EcoStore
26 Seafield Road
Hove, BN3 2TP
UK
Tel: 01273 245958
www.insightecostore.com

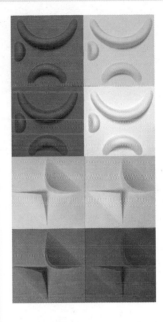

Mio Culture

Funky, eco-friendly design, with a range of furniture and wallpaper made of recycled products.
340 North 12th Street,
Unit 301
Philadelphia, PA 19107
USA
Tel: 215 925 9359
www.mioculture.com

Recycled Tableware

By Nature, a website for ethical living, does a small but fun range of recycled glass tableware including glasses, plates and platters.
By Nature Ltd
71 Avoca Road
London, SW17 8SL
UK
Tel: 0845 456 7689
www.bynature.co.uk

The Log Maker

This dinky machine recycles your newspapers and turns them into 'bricks' to burn on the fire.
The Green Shop
Cheltenham Road
Bisley
Stroud
Gloucestershire, GL6 7BX
UK
www.greenshop.co.uk

Materials

Tectan

An amazing new chipboard-like material made entirely from recycled beverage cartons that can then be refashioned into furniture, stationery and a plasterboard alternative. Based in Germany.

www.tectan.de

Glasstile

A broad range of lovely glass tiles for walls and floor, made of 85% recycled glass. They are available in USA, UK, Europe and Japan. Find your nearest dealer on:

www.glasstile.com

Smile Plastics

A British company that makes sheets of recycled plastic out of mobile phones, kids boots, bottles and CDs.

Mansion House
Ford
Shrewsbury, SY5 9LZ
UK
Tel: 01743 851 067
www.smile-plastics.co.uk

Vetrazzo

A material made of 85% recycled glass, used for countertops and as a non-structural building material. Pretty and colourful – beats laminate any day.

Counter/Production
710 Bancroft Way
Berkeley, CA 94710
USA
Tel: 510 843 6916
www.counterproduction.com

Straw Bale Building

Not strictly a material, but a website for people who are interested in building with straw bales – a sustainable building technique – with a thorough how-to section.

The Straw Bale Building Association
Hollinroyd Road
Butts Lane
Todmorden, OL14 8RJ
UK
Tel: 01442 825 421
www.strawbalebuildingassociation.org.uk

Versa-tile

A heavy-duty, hard wearing floor-tile made out of recycled plastic in a range of colours and for reasonable prices.

Versa-Tile
31 Bourne View Road
Netherton
Huddersfield
West Yorkshire, HD4 7JY
UK
Tel: 01484 663 142
www.versa-tile.co.uk

Environ

A biocomposite material made from recycled paper and soy flour. It looks a bit like granite and is 'harder than oakwood'. It can be used for flooring and furniture.

221 Mohr Drive
Mankato, MN 56001
USA
Tel: 800 324 8187
www.environbiocomposites.com

Eco-friendly flooring

Every type of flooring you could wish for – all made from recycled/eco-friendly materials including recycled aluminium, cork, glass and bamboo.

100 S Baldwin St, Suite 110
Madison, WI 53703
USA
Tel: 1 866 250 3273
www.ecofriendlyflooring.com

Yemm & Hart

This is a company that promotes three different types of recycled materials, with examples of products and installations by artists and designers using those materials.

1417 Madison 308
Marquand, MO 63655–9153
USA
Tel: 573 783 5434
www.yemmhart.com

Re-Design

Re-Design is a London based, not-for-profit organisation set up by Sarah and Jason Allcorn, to promote and showcase domestic designs that are friendly to the consumer, society and the environment. The view is that design can be a force for positive social change. In Sarah's words "Eco-designers are able to help make sustainability physically possible and aesthetically, intellectually and emotionally attractive."

In September 2005, Re-Design arranged an event exhibiting the work of 50 UK designers ranging from established, award winning companies to new, up-and-coming designers; all of which exemplified the concepts of sustainable, friendly and attractive design. Here are some of the highlights:

The U.R.E. Chair
A chair made of 100% mixed recycled plastics melted down and squeezed through a machine – thus eliminating large parts of the recycling process. Available directly via designer Richard Liddle.
Tel: 07957 400979
e-mail: richard@cohda.com
www.cohda.com

Prudence Floor Lamps

Made of recycled jumper sleeves, each light is different depending on the colour, style and density of the knit – and reasonably priced.

Available from
www.beefdesign.com
Or email the designer,
Lucy Turner directly:
beefdesign@hotmail.com

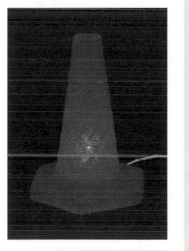

Shredded Paper Recycling Bin

A recycling bin made of the material it's intended for. Lightweight and durable, this funky accessory is made of a new material developed by its designer, Lynn Kingelin at the RCA.

Available via Mint
70 Wigmore Street
London, W1
Or through the designer
directly on:
www.ikuinen.com
or email directly:
lynnkingelin@ikuinen.com

Swirl Lamp

Made from a reclaimed spinning chimney cowl, the polished steel creates a fire-like effect which shimmers round the room as it spins. Currently made to order by the designer, it is available from:

Dan Davis-James Lighting
and Furniture Design
10 Burford Drive
Manchester, M16 8JF
Tel: 0161 861 7778
www.ddjdesign.co.uk

Me Old China

Brings new life to old crockery by building the odds and ends of dinner services into a range of two and three tier cake plates.

Tel: 07971 992258

www.peoplewillalwaysneedplates.co.uk

email: Hannah@peoplewillalwaysneedplates.co.uk

Sunday Papers

This clever chair is made out of tightly rolled newspapers bound together. The designer also makes tables and stools using the same technique.

www.stovelldesign.co.uk or email the designer: david@stovelldesign.co.uk

WEmake Street Sofa

Designed by one of the organisers of Re-Design, Jason Allcorn, this is a clever transformation of an old Euro-bin into a funky sofa, which is a lot more comfy than it looks.

WEmake

11 Woodlands

Clapham Common Northside

London, SW4 0RJ

www.wemake.co.uk

email: info@wemake.co.uk

Pli Chair

Made of 20 per cent post-consumer waste plastic, this chair is produced by a young company expressly manufacturing eco-furniture.
Pli Design
Unit D15, Parkhall
Road Trading Estate
62 Tritton Road
London, SE21 8DE
www.plidesign.co.uk
email: office@
plidesign.co.uk

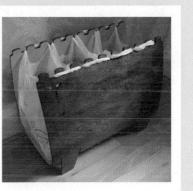

Sprout Binvention

Sprout are an eco-consultancy aiming to integrate sustainable design into mainstream practice. Their Binvention provides a creative solution to separating rubbish for recycling by reusing carrier bags.
Sprout Design Ltd
Studio One
Hoxton Works
128 Hoxton Street
London, N1 6SH
Tel: 0207 012 1713
www.sproutdesign.co.uk
email: info@sproutdesign.co.uk

Bath Chair

How better to recycle an old enamel bath? An exciting and inventive young designer, Tommy Allen is currently making these chairs to order.
4 St Mary's Gardens
London, SE11 4UD
Tel: 07815 108 950
email: allen.tommy@
gmail.com

Recycling directory of governmental organisations

If you are looking to find out about the practicalities of recycling in your area, your first port of call should be logging onto the website of your local authority. Due to space limitations, we are unable to list every single authority in this book. However, this section of the directory will list the national and state-wide contact details of environmental authorities, and their websites will often re-direct you to your local recycling authority.

UNITED KINGDOM

Accessing your local authority in the UK shouldn't be too hard. The WRAP and RecycleNow websites offer a postcode search to find your nearest recycling authority. This directory lists recycling information for the major UK cities. For all outside those cities, those two organisations should be your first port of call. DEFRA has detailed facts about recycling laws and programmes in the UK, and local websites like 'Lets Sort It Sheffield' and Gateshead Council

have easily accessible, basic information about recycling processes and contacts.

DEFRA

DEFRA (the Department for Environment, Food and Rural Affairs) works for the essentials of life – food, air, land, water, people, animals and plants. Its remit is the pursuit of sustainable development – weaving together economic, social and environmental concerns. DEFRA therefore:

- *Brings all aspects of the environment, rural matters, farming and food production together.*
- *Is a focal point for all rural policy, relating to people, the economy and the environment.*
- *Has roles in both European Union and global policy making, so that its work has a strong international dimension.*

Nobel House
17 Smith Square
London, SW1P 3JR
Tel: 08459 33 55 77
www.defra.gov.uk/
environment/waste/index

WRAP

The Waste and Resources Action Programme was set up in 2000 in response to the Government's new waste strategy to promote sustainable waste management. It is a not-for-profit organisation and works closely with DEFRA.
The Old Academy
21 Horse fair
Banbury
Oxon, OX16 0AH
Tel: 0808 100 20 40
www.wrap.org.uk

Recycle Now

A fun and accessible website and advertising campaign run by WRAP, pointing you in the direction of your local recycling facilities, and explaining recycling in easy bite-size nuggets.
Contact via WRAP or by e-mail on the website:
www.recyclenow.com

Capital Waste Facts

This site, run by the Greater London Authority, is a one-stop-shop for information and data on recycling and waste in London, enabling you to compare performance and view best practice across London. capitalwastefacts.com shows you how waste is being tackled and which recycling services are being provided in your area, as well as providing comprehensive waste and recycling data and information about innovations and new projects in London.

www.capitalwastefacts.com

Gateshead Council

Well laid out, accessible and informative.

Local Environmental Services
Park Road
Gateshead, NE8 3HN
Tel: 0191 433 7000
www.gateshead.gov.uk/environment/recycling.htm

Greater Manchester Waste Disposal Authority (GMWDA)

Medtia Chambers
5 Barn Street
Oldham, OL1 1LP
Tel: 0161 911 3581
www.gmwda.gov.uk

Birmingham Waste Minimisation Team

Environmental and Consumer Services Department
Montague Street Depot
Montague Street
Birmingham, B9 4BA
Tel: 0121 303 1935
www.birmingham.gov.uk/recycling

Let's Sort it Sheffield

Partly run by local authorities and partially independent – it's a great, well laid out website, and the people who run it are approachable and friendly.

Admail 3748
Sheffield, S1 1ZU
Tel: 0114 272 7000
www.letssortitsheffield.org

Bristol Recycling

Neighbourhood and Housing Services
Brunel House
St George's Road
Bristol, BS1 5UY
Tel: 0117 903 1221
www.bristol-city.gov/recycling

Cardiff Waste Forum

PO Box 645
Cardiff, CF24 4XJ
Tel: 029 2025 0030
www.cardiffwasteforum.org.uk

Plymouth Recycling Team

Department of Development
Plymouth City Council
Civic Centre
Plymouth, PL1 2AA
Tel: 01752 304 750
www.plymouth.gov.uk/homepage/environment/recycling

Waste Aware Scotland

Equivalent to the Recycle Now website for Scotland. It has a detailed area search to find your nearest local authority.

Islay House
Livilands Lane
Stirling, FK8 2BG
Tel: 01786 468 246
www.wascot.org.uk

Waste Awareness Wales

As above – with a postcode search to find your nearest local authority.

Local Government House
Drake Walk
Cardiff, CF10 4LG
Tel: 0845 330 5540
www.wasteawarenesswales.org.uk

USA

Environmental Protection Agency

This is the national governmental agency for all environmental issues including recycling in the US. It is a comprehensive source with links to other US based organisations and state-wide facilities.

Ariel Rios Building
1200 Pennsylvania Avenue
Washington, DC 20460
Tel: 202 272 0167
www.epa.gov

Alabama

Alabama Department of Environmental Management
PO Box 301463
Montgomery
AL 36130–1463
Tel: 334 271 7700
www.adem.state.al.us

Alaska

Department of Environmental Conservation
410 Willoughby
Avenue, Suite 303
Juneau, AK 99801–1795
Tel: 907 269 7802
www.dec.state.ak.us

Solid Waste Alaska Network
320 W Willoughby Avenue
Suite 300
Juneau, AK 99801
Tel: 1800 344 1432 ext 7184
www.ccthita-swan.org/main/recycling

Alaskans for Litter Prevention and Recycling
PO Box 200393
Anchorage, AK 99520
Tel: 907 274 3266
www.alparalaska.com

Recycling in Anchorage:
Anchorage Solid Waste Services
632 W 6th Avenue
Anchorage, AK 99501
www.muni.org/sws/recycling

Arizona

Arizona Department of Environmental Quality
Phoenix Main Office
1110 W Washington St
Tel: 800 234 5677
Phoenix, AZ 85007
www.azrecycles.gov

University of Arizona Recycling Council
Contact via website
recycle.web.arizona.edu

Arkansas

Arkansas Department of Environmental Quality
8001 National Drive
Little Rock, AR 72209
Tel: 501 682 0744
www.adeq.state.ar.us/solwaste/branch_recycling/default

California

Califorina Integrated Waste Management Board
Very comprehensive website.
1001 I Street
PO Box 4025
Sacramento
CA 95812–4025
Tel: 916 341 6000
www.recycle.ca.gov

Colorado

Colorado Department of Public Health and Environment
4300 Cherry Creek Drive South
Denver, CO 80246–1530
Tel: 303 692 2000
www.cdphe.state.co.us

Delaware

Delaware Department of Natural Resources and Environmental Control
89 Kings Hwy Dover
DE 19901
Tel: 302 739 9403
www.dnrec.state.de.us/DNREC2000/Recycling

Florida

Department of Environmental Protection
3900 Commonwealth Blvd
MS 49
Tallahassee, FL 32399
Tel: 850 245 2118
www.dep.state.fl.us/waste/categories/solid_waste

Georgia

**Department of
Natural Resources**
2 Martin Luther King,
Jr Drive, SE
Suite 1252, East Tower
Atlanta, GA 30334
Tel: 404 656 3500
www.gadnr.org/
or www.dca.state.ga.us
or www.p2ad.org/Assets/
Documents/ci_recysw

Idaho

**Department of
Environmental Quality**
1410 N. Hilton
Boise, ID 83706
Tel: 208 373 0502
www.deq.state.id.us/waste/
recycling/recycling.cfm

Illinois

**Illinois Recycle
Association (IRA)**
PO Box 3717
Oak Park, IL 60303–3717
Tel: 708 358 0050
www.illinoisrecycles.org

Indiana

**Indiana Department of
Environmental Management**
Indiana Government Center
North 100 N Senate Ave
Indianapolis, IN 46204
Tel: 800 451 6027 (toll-free
in Indiana) or 317 232 8603.
www.in.gov/idem/
oppta/recycling/

Iowa

**Iowa Department of
Natural Resources**
502 East 9th Street
Des Moines, IA 50319–0034
Tel: 515 281 8176
www.iowadnr.com/
waste/recycling/index

Kansas

**Bureau of Solid Waste
Management**
1000 SW Jackson Street
Topeka, KS 66612–1366
Tel: 785 296 1600
www.kdhe.state.
ks.us/kdsi/index

Kentucky

**KY Division of Waste
Management**
14 Reilly Road
Frankfort, KY 40601
Tel: 502 564 6716
www.waste.ky.gov/recycling

Louisiana

**Louisiana Department of
Environmental Quality**
602 North Fifth Street
Baton Rouge, LA 70802
Tel: 225 342 1234
www.deq.louisiana.gov/
assistance/recycling/index

Maine

**Department of
Environmental Protection**
17 State House Station
Augusta, ME 04333–0017
Tel: 207 287 7688
www.maine.gov/dep/
rwm/recycle/index

Maryland

**Maryland Department
of the Environment**

1800 Washington Blvd,
Baltimore, MD 21230
Tel: 410 537 3000 or
800 633 6101
www.mde.state.md.us/
Programs/LandPrograms/
Recycling/home/index.asp

Massachusetts

**Massachusetts Department
of Environmental Protection
– Recycling Section**

1 Winter Street
Boston, MA 02108–4746
Tel: 617 292 5500
www.mass.gov/dep/
recycle/recycle

Mississippi

**MS Dept. of
Environmental Quality**

Recycling and Solid Waste
Reduction Program
PO Box 10385
Jackson, MS 39289–0385
www.deq.state.ms.us

Minnesota

**Minnesota Pollution
Control Agency**

520 Lafayette Road
S. Paul, MN 55155–4194
Tel: 651 296 6300 or
800 657 3864
www.pca.state.mn.us/
backyard/index

Or in the twin cities:
www.greenguardian.com

Michigan

Michigan Recycling Coalition

3225 W St Joseph
Lansing, MI 48917
Tel: 517 327 9207
www.michiganrecycles.org

Missouri

**Department of Natural
Resources**

PO Box 176
Jefferson City, MO 65102
Tel: 1800 361 4827
www.dnr.state.mo.us/
regions/deqregions

Montana

**Department of
Environmental Quality**

Lee Metcalf Building
1520 E Sixth Avenue
PO Box 200901
Helena, MT 59620–0901
Tel: 406 444 2544
www.deq.state.mt.us/
Recycle/index.asp

Nebraska

**Nebraska Department of
Environmental Quality**

1200 "N" Street, Suite 400
PO Pox 98922
Lincoln, Nebraska 68509
Tel: 402 471 2186
www.deq.state.ne.us

Nevada

**Division of Environmental
Protection**

901 South Stewart Street
Suite 4001
Carson City, NV 89701–5249
Tel: 775 687 4670
ndep.nv.gov/recycl/recycle

New Hampshire

New Hampshire Department of Environmental Services
29 Hazen Drive
PO Box 95
Concord, NH 03302–0095
Tel: 603 271 3503
www.des.state.nh.us/
waste_intro

New Jersey

Department of Environmental Protection
401 E State Street
Trenton, NJ 08625
Tel: 609 984 6880
www.nj.gov/dep/dshw/
recycle/whyrecycle/index

New Mexico

New Mexico Environment Department
PO Box 26110
1190 St Francis Drive
Suite N4050
Santa Fe, NM 87502
Tel: 800 219 6157
www.nmenv.state.nm.us/
waste

New York

Department of Environmental Conservation
625 Broadway
Albany, NY 12233
Tel: 518 402 8549
www.dec.state.ny.us/website/
dshm/redrecy/index

NYC Waste Le$$
A comprehensive resource for recycling in New York City, run by the department of sanitation. Contact via e-mail only
www.nycwastless.org

North Carolina

Department of Environment and natural resources
1601 Mail Service Center
Raleigh, NC 27699–1601
Tel: 919 733 498
www.p2pays.org/
recycleguys/home.asp

Ohio

Office of Compliance Assistance and Pollution Prevention
Ohio Environmental Protection Agency
PO Box 1049
Columbus, OH 43216–1049
Tel: 614 644 3469
www.epa.state.oh.us/
ocapp/recycle

Oklahoma

Oklahoma Department of Environmental Quality
707 N Robinson
Oklahoma City, OK 73102
Tel: 405 702 1000
www.deq.state.ok.us/
lpdnew/recyclingindex

Oregon

Department of Environmental Quality, Solid Waste Policy and Program Development Section
811 SW Sixth Avenue
Portland, OR 97204,
Tel: 503 229 5913 or
800 452 4011
www.deq.state.or.us/wmc/
solwaste/recycling

Pennsylvania

Department of Environmental Protection
Bureau of waste management
Rachel Carson State
Office Building
400 Market Street
Harrisburg, PA 17105
Tel: 717 787 7382
www.dep.state.pa.us/dep/
deputate/airwaste/wm

WASTE IS A PROBLEM WE CAN DO SOMETHING ABOUT!
(LOS DESECHOS SON UN PROBLEMA DEL QUE PODEMOS HACER ALGO AL RESPECTO)

We Recycle In Our School
(En nuestra escuela ¡Reciclamos!)

Rhode Island

**Department of
Environmental Management**

235 Promenade Street
Providence, RI 02908–5767
Tel: 401 222 6800
www.dem.ri.gov/topics/citizen

South Carolina

**S.C. Department of Health
and Environmental Control**

Broad River Road
Columbia, SC 29210
Tel: 1800 768 7348
www.scdhec.net/recycle/index

South Dakota

**Department of Environment
and Natural Resources**

523 East Capitol Ave.
Pierre, SD 57501
Tel: 605 773 3153
www.state.sd.us/denr/des/
wastemgn/Recycling/
RecycleGuide.htm

Tennessee

**Department of Environment
and Conservation**

401 Church Street
L & C Annex, 1st Floor
Nashville, TN 37243–0435
Tel: 888 891 8332
www.state.tn.us/
environment/swm/prwr

Texas

**Texas Natural Resource
Conservation Commission**

12100 Park 35 Circle,
Austin, TX 78753
Tel: 512 239 1000
www.tnrcc.state.tx.us/
exec/sbea/rtol

Utah

**Department of
Environmental Quality**

150 North 1950 West
Salt Lake City, UT 84116
Tel: 801 538 6765
www.deq.utah.gov

Vermont

**Department of
Environmental Conservation**

One South Building
103 South Main Street
Waterbury VT, 05671–0401
Tel: 1800 932 7100
www.anr.state.vt.us/dec/
wastediv/R3/WReduct

Virginia

**Virginia Department of
Environmental Quality
Litter Prevention and
Recycling Program**

P.O. Box 10009
Richmond, VA 23240–0009
Tel: 804 698 4215
www.deq.state.va.us/recycle

Washington

**Washington State
Department of Ecology**

P.O. Box 47600
Olympia, WA 98504–7600
Tel: 1800 RECYCLE
www.1800recycle.wa.gov

West Virginia

**West Virginia Solid Waste
Management Board**
601, 57th Street
SE Charleston, WV 25304
Tel: 304 926 0448
www.state.wv.us/swmb

Wisconsin

**Department of
Natural Resources**
101 S Webster Street
PO Box 7921
Madison, WI 53707–7921
Tel: 608 266 2621
www.dnr.state.wi.us/org/
aw/wm/recycle/reference

Wyoming

**Department of
Environmental Quality**
122 West 25th St
Herschler Building
Cheyenne, WY 82002
Tel: 307 777 7937
www.trib.com/wyoming/recycle

CANADA

Natural Resources Canada
*A federal service specialising
in energy, minerals, metals,
forests and earth science.
With a good comprehensive
section on recycling.*
580 Booth Street
Ottawa, ON K1A 0E4
Tel: 613 947 3729
www.recycle.nrcan.gc.ca

ICLEI
*Local Governments
for Sustainability
– a democratically governed
association of cities, towns,
counties and local government
associations promoting
sustainability. It is worldwide,
but based in Canada.*
City Hall, West Tower,
16th Floor
100 Queen Street West
Toronto, ON M5H 2N2
Tel: 416 392 1462
www.iclei.org

AUSTRALIA

**Department of the
Environment and Heritage**
*This is the department in
charge of the government's
'No-Waste by 2010' strategy.*
GPO Box 787 Canberra
ACT 2601
Tel: 02 6274 1111
www.deh.gov.au

Act NOWaste
*A more user-friendly
website serving the above
department's No Waste
strategy with research,
publications and information
sheets available on-line.*
12 Wattle Street
Lyneham ACT 2602
Tel: 02 6207 2500
www.nowaste.act.gov.au

SOUTH AFRICA

**Department of
Environmental Affairs
and Tourism**
Address not listed, contact
via telephone or e-mail
Tel: 012 310 3911
www.environment.gov.za

NEW ZEALAND

Ministry of the Environment
Environment House
23 Kate Sheppard Place
Thorndon
PO Box 10362
Welllington
Tel: 04 439 7400
www.mfe.govt.nz

Reduce Your Rubbish
*A government initiative
with practical advice on
how to do exactly that.*
Ministry for the Evironment
PO Box 10362
Wellington
Tel: 04 439 7400
www.reducerubbish.govt.nz

EUROPE

AUSTRIA

**Federal Ministry for
Environment, Youth
and Family Affairs**
Stubenbastei 5
A–1010
Vienna
Tel: 1 515 22
www.bmu.gv.at/

**Alstoff Recycling
Austria (ARA)**
Neualmerstrasse 37
A–5400, Hallein
Tel: 664 53 27 179

BELGIUM

**Directorate General
Environment**
Place Victor Horta 40, Box 10
1060 Bruxelles
Tel: 02 524 95 00
www.environment.fgov.be

Electronics Recycling
www.recupel.be

DENMARK

**Environmental
Protection Agency**
Strandgade 29
1401 K, Copenhagen
Tel: 32 66 01 00
www.mim.dk

**City of Copenhagen
Recycling Services**
Kraftsværksvej 25
2300 København S
Tel: 70 10 18 98 or
32 66 18 98
www.kbhbase.kk.dk

GERMANY

Federal Ministry for the Environment
Postfach 12 06 29
53048, Bonn
Nordrhein-Westfalen
Fax: 1888 305 3225
www.bmu.de

Green Dot
Frankfurter Strasse 720–726
51145, Köln/Porz-Eil
Tel.: 0 22 03 9 37 0
www.dsd-ag.de

IT Recycling
Contact via website
www.recycle-it.de

IRELAND

Ministry for the Environment, Waste Management Section
Custom House, Dublin 1
Dublin
Tel: 353 1 679 3377
www.environ.ie

For links to your local recycling centre:
www.raceagainstwaste.com

ITALY

Ministero dell'ambiente
Via Della Ferratella
in Laterno 33
00184
Roma
Tel: 06 77 25 70 13

NETHERLANDS

Ministry of Housing, Spatial Planning and the Environment
Postbus 20945
2500 EZ, Den Haag
Tel: 070 339 39 39
www.vrom.nl

SPAIN

Ministry of the Environment
Plaza de San Juan de la Cruz
28071
Madrid
Tel: 91 597 6000
www.mma.es

General information about recycling:
www.redcicla.com

For kids:
www.educared.net/
concurso/586/reciclaje.htm

SWEDEN

Swedish Environmental Protection Agency
Blekhomsterassen 36
SE 106 48
Stockholm
Tel: 46 8 698 1000
www.naturvardsverket.se

SWITZERLAND

Swiss Recycling
Naglerwiesenstrasse 4
8049, Zürich
Tel: 044 342 20 00
www.swissrecycling.ch

International directory of non-governmental organisations (NGOs)

Greenpeace
Greenpeace is a non-profit organisation, with a presence in 40 countries across the world. As a global organisation, Greenpeace focuses on the most crucial worldwide threats to the planet's biodiversity and environment.
www.greenpeace.org/international

WWF
Estasblished in 1961, WWF operates in more than 100 countries, prioritising the areas representing globally outstanding examples of biodiversity. It attempts to tackle the social, economic and policy issues which are critical to sustainable livelihoods for people and the ecosystems upon which they depend.
www.panda.org

Friends of the Earth
An international network of grassroots groups in more than 70 countries, tackling a broad range of environmental issues.
www.foe.org

UK

Waste Watch
Probably the most comprehensive source on recycling in Britain, supplying detailed information sheets about materials, how to recycle at home and at the office and useful links to relevant organisations.
56–64 Leonard Street
London, EC2 4JX
Tel: 0207 549 0300
www.wasteonline.org.uk

Community Recycling Network (CRN)
Promotes community-based waste management.
Trelawney House
Surrey Street
Bristol, BS2 8PS
Tel: 0117 942 0142
www.crn.org.uk

Community Recycling Network, Wales (CYLCH)
Cardiff Business Technology Centre
Senghennydd Road
Cardiff, CF24 4AY
Tel: 029 2064 7000
www.cylch.org

Community Recycling Network Scotland
CRNS
Suite 27, Stirling Business Centre
Wellgreen Place
Stirling, FK8 2DZ
Tel: 01786 469 002
www.crns.org.uk

Friends of The Earth
26–28 Underwood Street
London, N1 7JQ
Tel: 0207 490 0881
www.foe.co.uk

GREENPEACE

Greenpeace
Canonbury Villas
London, N1 2PN
Tel: 0207 865 8100
www.greenpeace.org.uk

Oxfam
Runs the 'bring bring' scheme for recycling mobile phones as well as toner cartridges, computers, books and clothes.
Oxfam House
John Smith Drive
Oxford, OX4 2JY
Tel: 0870 333 2700
www.oxfam.org.uk

Groundwork
A charity for environmental regeneration, which works together with local trusts to improve the quality of regional environments.
85–87 Cornwall Street
Birmingham, B3 3BY
Tel: 0121 236 8565
www.groundwork.org.uk

Association of Charity Shops
Supports the fundraising initiatives of charity shops by pooling resources and expertise.
5th Floor, Central House
Upper Woburn Place
London, WC1H 0HE
Tel: 0207 255 4470
www.charityshops.org.uk

World in Sight
This is an initiative run by Help the Aged for recycling spectacles. The glasses you donate are sorted, cleaned and graded and delivered to ophthalmic clinics in developing countries. You can hand in your old glasses at any Dollond and Aitchison branch.
c/o Help the Aged
207–221 Pentonville Road
London, N1 9UZ
Tel: 0207 278 1114
www.helptheaged.org.uk

Global Action Plan
Designed to give practical advice on reducing energy and resource waste at home and at work.
8 Fulwood Place
London, WC1V 6HG
Tel: 0207 405 5633
www.globalactionplan.org.uk

L I L I

Low Impact Living Initiative
A charity whose mission is to help people reduce their impact on the environment.
Redfield Community
Winslow
Bucks, MK18 3LZ
Tel: 01296 714 184
www.lowimpact.org

Green Works
Recycling office furniture for the community. Donate or buy very reasonably online.
2nd Floor Downstream
1 London Bridge
London, SE1 9BG
Tel: 0845 230 2 231
www.green-works.co.uk

The Environment Council

An independent UK charity bringing together various skillsets to develop solutions to environmental issues.

212 High Holborn
London, WC1V 7BF
Tel: 0207 836 2626
www.the-environment-council.org.uk

Sustainable Communities Initiative

Builds towards Zero Waste communities.

Craigencalt Farm
Kinghorn
Fife, KY3 8YG
www.sci-scotland.org.uk

THE *WASTE*BOOK

The Wastebook

A guide to recycling in London and South East England.

www.recycle.mcmail.com/

THE RECYCLING CONSORTIUM

The Recycling Consortium (TRC)

Create Centre
Smeaton Road
Bristol, BS1 6XN
Tel: 0117 930 4355
www.recyclingconsortium.org.uk

Let's Recycle

Let's Recycle is the home of news and information for recyclers and all those involved in sustainable waste management in the UK. It is the UK's only independent dedicated website for businesses, local government and community groups involved in recycling and waste management. It delivers news and material prices as well as key information for the business sector. Some information is provided for educational purposes.

3 Downstream
1 London Bridge
London, SE1 9BG
Tel: 0207 785 6443
www.letsrecycle.com

Recycle More

General information and chatgroups. Fairly basic, but good for kids. Run by Valpak, the UK Producer Compliance Scheme.

Valpak Ltd
Stratford Business Park
Banbury Road
Stratford Upon Avon
CV37 7GW
Tel: 08450 682 572
www.recycle-more.co.uk

Encams

This is a charity for environmental campaigning best known for their 'Keep Britain Tidy' campaign. It is partially government funded, and focuses more on the problems of litter, flyposting and urban cleanliness than on issues like recycling. It's good for schools and educational organisations.

Elizabeth House
The Pier
Wigan, WN3 5 FX
Tel: 01942 612 621
www.encams.org

Valpak

Not an NGO as such, but a leading provider of producer responsibility and recycling solutions for UK businesses. Lots of interesting info about new recycling technologies too!

Tel: 01789 298 769
www.valpak.co.uk

UNITED STATES

Earth 911
A very comprehensive website covering all environmental issues. It has a postcode insert box, to direct you to your local recycling centre. Their slogan is 'make every day an earth-day'.
7301 E. Helm, Building D
Scottsdale, AZ 85260
Tel: 480 889 2650
www.earth911.org

National Resource Defense Council (NRDC)
A general nature protection website, with information about recycling and a guide to greener living.
40 West 20th Street
New York, NY 10011
Tel: 212 727 2700
www.nrdc.org

Grassroots Recycling Network
A community based organisation aiming for Zero Waste.
4200 Park Blvd Suite 290
Oakland, CA 94602
Tel: 510 531 5523
www.grrn.org

Northeast Recycling Council
An organisation for ten northeastern states to combine forces and develop recycling markets.
139 Main Street Suite 401
Brattleboro, VT 05301
Tel: 802 254 3636
www.nerc.org

National Recycling Coalition
A membership organisation of recycling professionals to maximise recycling – includes where and how to recycle.
1325 G Street NW
Suite 1025
Washington, DC 20005
Tel: 202 347 0450
www.nrc-recycle.org

New American Dream
Helping Americans consume responsibly and live consciously. This organisation aims to protect the environment, to enhance quality of life and to promote social justice.
6930 Carroll Ave, Suite 900
Takoma Park, MD 20912
Tel: 301 891 3683 or
1 877 68 DREAM
www.newdream.org

Environmental Defense
Finding innovative, practical ways to solve the most urgent environmental problems.
257 Park Avenue South
New York, NW 10010
Tel: 212 505 2100
www.environmentaldefense.org

Reusablebags.com
A grassroots organisation aiming to raise plastic bag awareness, to offer viable alternatives and to enable people to consume less. They have a variety of calico and other bags available at their online shop, and useful articles about plastic bag solutions from around the world.
2121 W Division Street
Suite 3W
Chicago, IL 60622
Tel: 773 704 3421
www.reusablebags.com

Rainforest Alliance
Aims to protect ecosystems and the people and wildlife that depend on them.
665 Broadway, Suite 500
New York, NY 10012
Tel: 212 677 1900 or
1 888 MY EARTH
www.rainforest-alliance.org

The Green Scene
Environmental tips for radio including a fair amount on recycling.
Contact via e-mail on website
www.thegreenscene.com

Recycle.net

A great resource for individuals and industry professionals.
PO Box 1910
Richfield Springs, NY 13439
No phone, contact via e-mail
www.recycle.net

San Francisco Clean City

A community based website for keeping San Francisco green.
1016 Howard Street
San Francisco, CA 94103
Tel: 415 552 9201
www.sfcleancity.com

We Can

Promoting recycling of cans and bottles to help the homeless.
630 Ninth Avenue, Suite 900
New York, NY 10036
Tel: 212 262 2222
www.wecanny.org

Association of Vermont Recyclers

A non-profit association specialising in recycling and environmental education in Vermont.
PO Box 1244
Montpelier, VT 05601
Tel: 802 229 1833
www.vtrecyclers.org

Recycle This!

An activist run, New York based organisation for recycling.
Tel: 212 592 4184
www.recyclethisnyc.org

Brevard County Freecycle Network

A worldwide network dedicated to sharing community resources.
4633 Bluejay Lane
Melbourne, FL 32935
Tel: 321 757 9971
www.brevardfreecycle.tripod.com

Recycling Advocates

Formed in 1987, Recycling Advocates is a citizen-based, grassroots group dedicated to creating a sustainable future through local efforts to reduce, reuse and recycle.
P.O. Box 6736
Portland, OR
97228–6736
Tel: 503 777 0909
www.recyclingadvocates.org

CANADA

Pitch In Canada

A national NGO for the preservation of the environment including anti-littering and recycling information.
Box 45011, Ocean Park RPO
White Rock, BC V4A 9L1
Tel: 1 604 290 0498
www.pitch-in.ca

Association of Municipal Recycling Coordinators

An NGO set up to facilitate the sharing of municipal waste reduction and recycling information.
Suite 100,
127 Wyndham Street North
Guelph, ON N1H 4E9
Tel: 519 823 1990
www.amrc.guelph.org

Municipal Waste Integration Network (M-WIN)

A resource for municipal waste minimisation and management.
PO Box 1116
704 Glen Morris Road
West Ayr, ON N0B 1E0
www.mwin.org

AUSTRALIA

Clean Up Australia

A community based organisation for a cleaner Australia. Comprehensive website, good factsheets.
18 Bridge Road
Glebe, NSW 2037
Tel: 02 9552 6177
www.cleanup.com.au

Australian Conservation Foundation (ACF)

Floor 1, 60 Leicester Street
Carlton
Victoria, 3053
Tel: 03 9345 1111
www.acfonline.org.au

Green Innovations

An environmental think tank aiming for global and local sustainability.
195 Wingrove Street
Fairfield
Victoria, 3078
Tel: 03 9486 4799
www.green-innovations.asn.au

Greening Australia

Restoring health, diversity and productivity to the environment.
PO Box 74
Yarralumla ACT 2600
www.greeningaustralia.org.au

Urban Ecology Australia

Working to create ecologically integrated human settlements.
105 Sturt Street
Adelaide 5000
Tel: 08 8212 6760
www.urbanecology.org.au

SOUTH AFRICA

National Recycling Forum (NRF)

An NGO supporting the recycling industries in South Africa.
PO Box 79
Allen's Nek, 1737
Tel: 011 675 3462
www.recycling.co.za

Fairest Cape

Aimed at increasing waste awareness in Cape Town.
PO Box 97
Cape Town 8000
Tel: 021 462 2040
www.fairestcape.co.za

Natal Recycling Forum

PO Box 1535
Durban 4000
Tel: 031 376 243

National Coordinating Committee for Recycling

PO Box 1378
Pinegowrie 2123
Tel: 011 789 1101

NEW ZEALAND

ZEROWASTE
NEW ZEALAND TRUST

Zero Waste New Zealand Trust

PO Box 33 1695
Takapuna
Auckland
Tel: 64 9 486 0734
www.zerowaste.co.nz

Waste Management Institute of New Zealand (WasteMINZ)

A not-for-profit organisation promoting good waste management.
166 Kitchener Road
Milford
Auckland
Tel: 64 9 486 6722
www.wasteminz.org.nz

Business Care

Promoting cleaner production and waste minimisation in the working environment.

Contact via website
www.businesscare.org.nz

Environmental Choice

An organisation for eco-labelling and eco-friendly design.

PO Box 56 533
Dominion Road
Mt Eden
Auckland 1003
Tel: 64 9 845 3330
www.enviro-choice.org.nz

EUROPE

Rreuse

European Community Recycling Network

40 Rue Washington
B–1050 Brussels
Belgium
Tel: 32 2 647 9995
www.rreuse.org

Glossary

A

abatement – The reduction in landfill contamination through waste reduction and recycling.

aeration – The process of allowing circulating air to reach bulk material, such as compost.

agricultural waste – Waste from premises used for agriculture, including animal and plant waste resulting from the production and processing of farm or agricultural products, discarded pesticide containers, plastics, tyres, batteries, clinical waste, old machinery and oil.

aluminium can – A food or beverage container that is made of at least 94 per cent aluminium.

air emissions – Solid, gaseous or odorous substances discharged into the air. These can result from internal combustion engine exhausts, incineration, landfill, decomposition, demolition etc..

arability – The degree to which land can be cultivated for the planting and growing of crops.

B

backyard composting – The controlled biodegradation of materials on the site where they were generated, such as grass clippings, leaves, and/or other organic waste from the household.

Bakelite – A trademark used for any of a group of synthetic resins and plastics found in an assortment of manufactured products.

bale – A compacted and bound cube of recyclable material, such as waste paper, scrap metal, or rags.

biodegradable – A material that will decay as a result of the action of microorganisms breaking down the material into elements such as carbon that are recycled naturally.

biodiversity – The variability among living organisms on earth, including the variability within and between ecological communities or systems.

biomimicry – Design and manufacturing values and practices that imitate natural materials or processes.

C

centralised composting – A system of removal of waste to a central facility within a politically defined area with the purpose of composting garden waste.

CFC (chlorofluorocarbons) – A gas formerly used as a refrigerant and aerosol propellant containing carbon, hydrogen, chlorine, and fluorine. The chlorine in CFCs causes depletion of atmospheric ozone.

closed-loop recycling – The process in which something is recycled back into the same item, such as used aluminium cans being made into new cans.

compost – A relatively stable mixture of decayed plants and other organic materials used to enrich soil.

commercial waste (trade waste) – Waste arising from premises which are used wholly or mainly for trade, business, sport, recreation or entertainment, excluding household and industrial waste.

copper cementation – The process using troughs containing iron employed to recover copper otherwise lost in mine waters. The precipitated copper sinks to the bottom of the troughs as a sludgy mass of fine particles. It is removed periodically, dried, and transported to the smelter.

cornstarch plastic – A material composed of starch prepared from corn grains containing low to no levels of petroleum. Used increasingly as an alternative for making compostable, degradable bottles.

D

deforestation – The process of destroying forested areas and replacing it with something else, often as a result of human activities.

degradable plastic – A material capable of being molded, extruded, or cast into various shapes and forms with particular function of decomposition. The use of degradable plastics include food and beverage containers, packaging materials, golf tees, raincoats, nappies, and women's hygiene goods.

de-inking – The removal of ink from printed waste paper to allow reuse.

dioxins – Any of several carcinogenic organic chemical compounds, formally known as polychlorinated dibenzo-p-dioxins, that occur as impurities in petroleum derived herbicides.

downcycling – The recycling of a material into another material of a lesser quality.

durability – The extent to which a material, product or system lasts without sustaining damage or wear.

E

ecosystem – A contained group of mutually dependent organisms together with its environment, functioning as a unit.

effluents – Liquid waste discharged from a factory, nuclear plant or sewage system.

e-waste – A generic term for waste generated through discarded electrical products or devices that become obsolete, such as mobile phones, computers, or household appliances. Lead, cadmium, mercury and other toxic materials used in this equipment can contaminate the environment.

emission – A substance discharged into the air, especially by an internal combustion engine.

energy recovery – The conversion of solid waste into energy or saleable fuel.

F

feedstock recycling – The process of recycling raw materials used in the industrial manufacture of a product.

ferrous metals – Those metals which include iron and all iron derivatives naturally attracted to magnets, including steel, nickel, cobalt, neodymium, and praseodymium.

furnace – An enclosure in which energy in a non-thermal form is converted to heat, through a process of combustion, for example, to smelt metal.

fibre – The fine threads of natural or synthetic material which combine to form paper.

G

global economy – The international stretch of capitalism across national boundaries regulated only with minor restrictions.

grade – A categorisation of recycled products that separates them by material content or previous use.

granulator – A mechanical device used for the formation of small grains or particles of plastic.

greenhouse effect – The phenomenon whereby atmospheric pollution by gases traps solar radiation causing warming of the Earth's surface.

H

habitat – The natural conditions and environment in which an organism or ecological community lives.

HDPE (high-density polyethylene) – A recyclable plastic, used for milk and detergent containers, detergent containers, and base cups of plastic soft drink bottles.

high-grade paper – Often referring to good quality paper formed from virgin pulp – i.e. fibres that have not been recycled.

household waste – Waste generated in the home including food scraps, packaging, pet litter and nappies.

HHW (household hazardous waste) – A substance, or substances, used within the home that are potentially harmful to the environment, humans and other living organisms. These include home cleaning products and chemicals used for gardening.

I

incineration – The means of destroying waste by burning it in a furnace.

incinerator ash – Solid particles of carbon as a result of the burning of waste material.

industrial waste – Solid or liquid waste originating from the large-scale manufacture of materials, goods or products. Often used to describe types of waste with a low rate of composition or of a toxic nature.

inorganic – Used to describe substances composed of minerals rather than living matter.

J

Jazz PET – Coloured polyethylene terephthalate, often used for soft drink bottles, with a lesser post use value than natural PET on account of its discolouration when recycled. Opportunities to use Jazz PET in guttering and drainage applications, and potentially in coat hangers, are being explored.

K

Kyoto Protocol –
An agreement on global warming reached by the United Nations Conference on Climate Change in Kyoto, Japan, in 1997. The major industrial nations vowed to reduce their greenhouse gas emissions between 2008 and 2012, although the United States Senate has since refused to endorse the agreement.

L

landfill – The disposal or site of disposal of solid waste by burying it between layers of dirt in low-lying ground or excavated holes.

LDPE (low-density polyethylene) – A lightweight thermoplastic with a less dense molecular structure than that of high-density polyethylene.

logging – The work or business of felling, trimming and transporting trees for timber.

low-grade paper –
A generic term for all lesser quality paper, often used in newspaper publication, composed of recycled fibres from used paper.

M

man-made materials – Those substances made by human beings which do not occur naturally.

mechanical separation –
The process of sorting materials according to their composition through the utilisation of mechanical equipment, such as conveyer belts and magnets for the sorting of steel and aluminium cans.

methane – An odourless, colourless, flammable gas that is the major element of natural gas and is used as a fuel.

mulch – A protective covering of organic matter laid around plants to prevent erosion, retain moisture, and sometimes improve the properties of the soil.

N

natural resource – A material source of wealth, such as timber, fresh water, or coal that is naturally occurring and can be exploited by people.

non-biodegradable – Made of substances that will not decay and are not recycled naturally, such as glass, plastics and most synthetic fibres.

non-ferrous metals –
Metals that do not contain iron, for example, aluminium, brass and copper (which can remembered as ABC).

non-recyclable – The state of continuing existence or duration of a material whereby it cannot be reused or reconstituted to form another product or substance.

non-renewable resources –
Any resource, such as mineral reserves, that cannot or will not be replenished naturally in the course of time or as a result of over use by people.

O

organic – Relating to or derived from living organisms

organic waste – Waste composed of living matter, such as leaves, plants, and food scraps which have the properties to decay and return to nature.

P

PET (polyethylene terephthalate) – A plastic resin used to make food, beverage and other liquid containers. It is also one of the most important raw materials used in man-made fibres.

pit mining – An excavation or cut made at the surface of the ground for the purpose of coal mining and extracting metal ores, such as copper, gold, iron, and aluminium.

plastics – The broad term given to any of a group of extremely versatile synthetic materials made from the polymerization of organic compounds, which are capable of being molded, cast into various shapes and films, or drawn into filaments used as textile fibres.

pulp – A mixture of cellulose material, such as wood, paper, and rags, ground up and moistened to make paper.

pollutant – Something that causes harm to an area of the natural environment, for example, chemicals or waste products that contaminate the air, soil, or water.

PP (polypropylene) – A hard and tough thermoplastic resin used in making pipes, industrial fibres, and molded objects.

PS (polystyrene) – A rigid clear synthetic polymer of styrene that can be molded into objects or made into a foam that is used for insulation and packaging.

PVC (polyvinyl chloride) – A common, hard-wearing thermoplastic resin, used in a wide variety of manufactured products, including rainwear, garden hoses, phonograph records, and floor tiles.

R

radioactive – Used to describe a substance such as uranium or plutonium that emanates energy in the form of streams of particles, causes by the decaying of its unstable atoms.

raw materials – A unprocessed natural material that is used in manufacture.

recovery – Generating value from wastes from a variety of activities such as recycling, composting and energy recovery.

recycle – To adapt or process used or waste material so that it can be converted for a new use or used again for the same purpose.

renewable resources – A natural resource such as lumber that can be renewed as quickly as it is used up, or source of energy, for example solar, wind or tidal power, that can, in theory, be used to generate power indefinitely – unlike mineral resources.

retread – The process of fitting a new tread to a worn tyre.

S

scrap – Discarded waste material, especially metal suitable for reprocessing.

shoddy – The fibrous material that results from shredding textiles prior to recycling. It is divided into grades according to quality.

shredder – A machine that tears objects into smaller pieces allowing for greater ease in which to reconstitute materials into products.

sludge – A semisolid material composed of the remnants of the paper making process, including staples, ink, glue and small fibres.

solid waste – Discarded materials other than fluids, for example, food residues, yard trimmings, textiles, plastics, and paper.

surface mining – A type of mining used to extract deposits of mineral resources that are close to the surface. Surface mining generally leaves large devastated areas which has a huge negative effect on the local ecosystem and the environment.

T

Tetrapak™ – A Swedish food packaging company whose plastic coated paper cartons are used for storing and transporting milk, soup, fruit juices and other liquid products.

toxic – Relating to or containing a poison or toxin capable of causing serious harm or death, especially by chemical means.

U

upcycling – A process which involves improving the quality or function of a product after it is recycled.

V

vermicomposting – The use of specially bred worms (such as the Brandling worm, also known as the Tiger worm or Red Wriggler) to convert organic matter into compost.

virgin materials – New materials or those which have not yet been recycled. Used materials have to compete in the marketplace with new (virgin) materials as the cost of collecting and sorting the materials usually means that they are equally or more expensive than virgin materials.

vulcanised rubber – Strengthened rubber resulting from the industrial process of combining it with sulphur and other additives and then applying great heat and pressure.

W

waste – An unwanted or unusable substance or material.

waste management – The collection, transportation, processing or disposal of waste materials, in attempt to reduce their effect on human health or local amenity. Relatively recently, the focus has broadened to consider waste materials' effect on the environment.

waste reduction – The concerted effort by groups or individuals to minimise the amount of waste they or their companies produce, either as part of an environmental or economic strategy (or both).

white goods – Large household appliances, such as refrigerators, ovens and dishwashers, typically finished with white enamel, but now often coloured.

Selected bibliography

Brooks Paul, *The Pursuit of Wilderness*, Houghton Mifflin, 1971

Brown, Lester R, *Eco-Economy*, New York: W W Norton & Company, 2001

Browner, Michael and Warren Leon, *The Consumers Guide to Effective Environmental Choices*, New York: Three Rivers Press, 1999

Bruges James, *The Little Earth Book*, Disinformation Ltd, 2004

Carson, Rachel, *Silent Spring*, New York: Mariner Books, 2002

Christensen Karen, *The Armchair Environmentalist*, London: MQ Publications, 2004

Cothran, Helen, *Garbage and Recycling: Opposing Viewpoints*, Greenhaven Press, 2002

Friends of the Earth, *Save Cash and Save the Planet*, London: Harper Collins, 2005

Fuad-Luke, Alistair, *The Eco Design Handbook*, London: Thames and Hudson, 2005

Girling, Richard, *Rubbish!*, London: Eden Project Books, 2005

Gore, Al, *Earth in the Balance*, Plume Publishing ,1993

Harmonious Technologies, *Backyard Composting: Your Complete Guide to Recycling Yard Clippings*, Harmonious Technologies, 1995

Hawken, Paul, Amory B Lovins, L Hunter, *Natural Capitalism*, London: Earthscan Publications Ltd, 2000

Hayes, Denis, *The Official Earth Day Guide to Planet Repair*, Washington DC: Island Press, 2000

Hegarty, Mark and Neil Bennett, *The Little Book of Living Green*, London: Nightingale Press, 2000

Hill, Julia Butterfly, *One Makes the Difference, London:* Harper Collins, 2002

Langholz, Jeffrey and Kelly Turner, *You Can Prevent Global Warming and Save Money*, London: Andrews McMeel Publishing, 2003

Lilienfeld, Robert and William Rathje, *Use Less Stuff*, New York: Fawcett Books, 1998

Lund, Herbert F, *McGraw Hill Recycling Handbook*, New York: McGraw Hill, 2000

Mau, Bruce, *Massive Change*, London: Phaidon, 2004

McDonough, William and Michael Braungart, *Cradle to Cradle*, New York: North Point Press, 2002

Ortega y Gassett, Jose, *Meditations on Quixote*, Illinois: University of Illinois Press, 2000

Palmer, Joy A, *Fifty Key Thinkers on the Environment*, Routledge 2001

Papanek, Victor, *The Green Imperative*, London: Thames and Hudson, 2003

Pears, Pauline, *All About Compost*, London: Search Press, 1999

Ross, Simon and Kerski Joseph, *The Environment*, New York: Hodder Arnold, 2005

Ryan, John C, *Stuff*, Seattle: Northwest Environment Watch, 1997

Siegle, Lucy, *Green Living in the Urban Jungle*, London: Green Books, 2001

Smith, Paul Slee, *Recycling Waste*, Shrewsbury: Scientific Publications Ltd, 1976

Articles

Andalo, Debbie, "Council Makes Recycling Compulsory", *The Guardian*, 6 January 2005

Bloom, Sara, "How is Germany Dealing with its Packaging Waste?", *Whole Earth*, Winter 2002

Brockes, Emma, "Plastic Planet", *The Guardian*, 17 October 2002

Das, Sushi, "Dirty Old Bags", *The Age*, 29 June 2004

Donnelly, James E, "Numbers Never Lie, But What Do They Say? A comparative look at municpal solid waste recycling in the United States and Germany", *Georgetown International Environmental Law Review*, Fall 2002

Flanagan, Ben, "Turn Your Baby Green", *Observer*, 22 May 2005

Goldstein, Nora, "Comparing Composting Technologies at Correctional Facilities", *BioCycle*, March 2003 pp 28–32

Koike, Yuriko, "The Spirit of Mottai Nai", *Our Planet*, vol 16 no 1, UNEP Publications

Lerner, Jaime, "Change Comes From the Cities", *Our Planet*, vol 8.1, June 1996, UNEP Publications

Marion, Jim, "Composting 12,000 Tons of Food Residuals a Year", *BioCycle*, May 2000, pp. 30–35

Marion, James I, Correctional "System Wins with Composting and Recycling", *BioCycle*, September 1994, pp. 31–49

Meadows, Donella, "The City of First Priorities – Curitiba, Brazil", *Whole Earth Review*, Spring 1995

Schreurs, Miranda A, "Divergent Paths: Environmental Policy in Germany, the United States, and Japan", *Environment*, October 2003

Siegle Lucy, "Hell's Kitchen", *Observer Magazine*, 12 September 2004

Spencer, Robert, "Integrated Recycling Pays Off at Prison Facilities", *BioCycle*, May 1991 pp. 47–49

Suzuki, Shunichi, "Slimming the Waste", *Our Planet*, vol 16 no 1 UNEP Publications

Uela, Kazuhiro, "Reducing Household Waste: Japan learns from Germany", *Environment*, November 2001

Vidal, John, "Drowning in a Tide of Discarded Packaging", *The Guardian*, 9 March 2002

Web resources

Alucan, Aluminium Packaging Recycling Association
www.alupro.org.uk

Battery University
www.batteryuniversity.com

BBC News, "NI Shoppers Would Bring Their Own Bags"
www.new.bbc.co.uk/1/hi/northern_ireland/1854239.stm

BBC News, "Irish Bag Tax Hailed Success"
www.news.bbc.co.uk/1/hi/world/Europe/2205419.stm

British Glass
www.britglass.org.uk

Bureau of International Recycling (BIR)
www.bir.org

CBBC, "Make a Wormery"
www.bbc.co.uk/cbbc/xchange/doit/nature/article_make_wormery

Department of Trade and Industry, "Tyre Recycling Information Sheet"
www.dti.gov.uk

Envirogreen
www.envorgreen.co.uk/services_battery

Environment and Plastics Industry Council (EPIC)
www.cpia.ca/epic/media

EnvoCare
www.envocare.co.uk

Friends of the Earth, United Kingdom
www.foe.co.uk

Friends of the Earth, Paper Recycling: Exposing the Myths
www.foe.co.uk/resource/briefings/paper_recycling

Gateshead Council
www.gateshead.gov.uk

The Glass Packaging Institute
www.gpi.org

Green Dot Germany
www.dsd-ag.de

Greenpeace
www.greenpeace.org

How to Compost
www.howtocompost.org

Kawischan, Lupong, "Recycling Newspapers" CSM Earthworks Club
www.mines.edu/stu_life/organ/earth/newspaperfl

Let's Recycle
www.letsrecycle.com

Environment Agency, Life Cycle Assessment of Disposable and Reusable Nappies in the UK
www.environment-agency.gov.uk

Make-Stuff.com
www.make-stuff.com/recycling/paper

National Recycling Coalition
www.nrc-recycle.org

Newspaper Industry Environmental Technology Initiative
www.newspaper.paisley.ac.uk

Nova, Science in the News, "Making Packaging Greener – Biodegradable Plastics"
www.science.org.au/nova/061/061key.htm

Packaging Today
www.packagingtoday.com/introenvironment

Planet Ark
www.planetark.com

Recoup Recycling
www.recoup.org

The Recycling Consortium,
www.recyclingconsortium.org.uk

Saskatchewan Waste Reduction Council
www.saskwastereduction.ca/metal/steel

The Steel Recycling Institute
www.recycle-steel.org

WasteCap, Information on Recycling Steel Products
www.wastecap.org/wastecap/commodities/steel/steel

Waste Care
www.wastecare.co.uk/battery_recycling

Composting Council of Canada
www.compost.org/natural.html

Spokane Solid Waste
www.solidwaste.org

Waste Online information sheets
www.wasteonline.org.uk

Women's Environmental Network, "Environment Agency Nappy Report is Seriously Flawed"
www.wen.org.uk/general_ages/newsitems/ms_LCA19.5.05

WRAP
www.wrap.org.uk

Directory index

Index

Picture credits

Acknowledgements

Many people have been very forthcoming and generous with their time and effort. *Recycle, The Essential Guide* would not have been possible without them.

Many thanks to:

Alex Bratt who helped with picture research and wrote up Case studies.

Sion Parkinson, who provided invaluable research assistance and took a lot of the photos illustrated in this book.

Draught Associates.

Thanks also to:

Jim Marion from New York DOCS for his help on the Case study and the photographs accompanying it.

Megan Yates.

Robin Nicholson at Edward Cullinans Architects for his comments and feedback.

Jamie Anley at Jam.

Sarah Davis.

Angelina Li, who helped with early research.

Sura Malocco at Subranded.

Doina Petrecscou for sending us the text and imagery on the Dakar Case study.

Cam Mattheson at the Nightingale Estate in Hackney.

Ceridwen Johnson at Traid.

Kjerti Berg and Kari lill Ljøstad at Returkartong.

Everyone at Greenpeace UK, especially Daphne Christelis and Angela Glieneke who were so generous and helpful.

All the designers in our design directory who supplied imagery and information.

Laura Jansen at WEN.

A special thanks to Steve Haugh at Wastewatch and his colleagues at Recycle Now, who provided so many images used in this book.

Printed on Rolland Enviro100 80 lb. Text and 100 lb. Cover

Black Dog Publishing

Architecture Art Design Fashion History
Photography Theory and Things

© 2006 Black Dog Publishing Limited,
the artists and authors
All rights reserved

Edited by Duncan McCorquodale and Cigalle Hanaor
Introduction by Lucy Siegle
Text by Cigalle Hanaor
Designed by Draught Associates
Research and additional text by Alex Bratt and Sion Parkinson

Black Dog Publishing Limited
Unit 4.4 Tea Building
56 Shoreditch High Street
London E1 6JJ

Tel: 44 (0)20 7613 1922
Fax: 44 (0)20 7613 1944
Email: info@bdp.demon.co.uk

www.bdpworld.com

All opinions expressed within this publication are those
of the authors and not necessarily of the publisher.

British Library Cataloguing-in-Publication Data.

A CIP record for this book is available from the British Library.

ISBN 1 904772 36 6

Every effort has been made to trace the copyright
holders, but if any have been inadvertently overlooked
the publishers will be pleased to make the necessary
arrangements at the first opportunity.

Printed in Canada